It's Time:
Book 1: John Carpenter Speaks Out

BY
ALEXANDRIA ALTMAN

It's Time: John Carpenter Speaks Out

Library of Congress Cataloging in Publication Data

Published by:

Arrowhead Classics Publishers
P.O. Box 4189
Sevierville, TN 37864
Printed in USA

Cover Design: Alan Lyle

ISBN No. 1-885640-39-0

Dedication

John and Karen Carpenter and their family; my daughters Nicole and Tessa, who mean the world to me; my family who raised me; Mike and Belinda Berry for help and encouragement; and the memory of my grandmother, Lula Pearl Lamb, who I know watches over me - I love you.

Acknowledgements

Frank Bean of Coca-Cola, many thanks; Mike and Matt Berry, my two cute nephews, Bill East of Arrowhead Classics, a real sweetheart; Brenton Logan Gill, who owns my heart; Thomas Glisson, attorney and friend; Ed Johncox, scriptwriter and great friend; Patty and Doug Lambert, love always; Joe Meyers, the simplest millionaire I know; Barbra Nyenhovis, I will never forget you; Josh Powers, a little boy who I love like my own; Ron Smith, of Smith and Wesson, thanks for all your love and support and help; Dee Vinson, who always stood by me; and, of course, the Good Lord, without Whom there is nothing.

A very special thank you to the editor, Paul Farr.

Editor's Note

Like most of Alexandria Altman's work, this small book carries more astonishing punch than most volumes many times its size. If you are new to these topics, you've picked the perfect place to begin. If you've read about these subjects before, you'll find yourself understanding them in surprising and enlightening ways.

John F. Carpenter is truly one of the heroes of our time, both in the service of his country and in the service of physical and metaphysical discovery. Approaching life with a rare openness of mind and honesty of spirit, he embodies the ideal of the new renaissance man. Refusing to accept old ideas simply on faith, he follows the maxim of Sir Arthur Conan Doyle's famed investigator: First, eliminate the impossible, whatever is left, no matter how improbable, must be the truth.

Decide for yourself. But whether you agree or disagree, there can hardly by any doubt that the issues discussed by Alexandria and John in that tiny cafe are among the most important of our time.

Table of Contents

Chapter One

Meeting John Carpenter

The Appalachians are the oldest mountains on the Earth. Which means, I suppose, that they've seen the most, felt the most, heard the most, probably even forgot the most of just about any other landscape open to the air. Maybe only the oceans know more and hold more than these admittedly humble hills, all worn down and bowed over and mossy as they are, like some ancient congregation at the final sermon of Nature. But I'm here to tell you, they're not sleeping through that sermon. No, they are wide awake and listening, and they are singing with their still, small voices – voices that can't be heard over our car engines or our headsets, or even our own chewing. Everything has to stop. Only then you can hear them.

What they do is tell stories. They gossip about the old times, and tell tales they heard from the dead ones, and wonder at all the new things. *The world is changing so we don't even know it,* they say. *But we are old and what can we do? Let it change, then. We'll be gone soon.* That's what you hear them say when everything is quiet.

But when the sun is shining and the day is cool, and driving through the mountains is like shooting rapids of brilliant colored gemstones in a comfortable canoe, you don't think about the world like that. You've got a different, younger way of seeing things for the moment, and it suits you fine. In fact, that's why you're out here, to get that feeling again, and the hills give it to you gladly. That's what Ron and I used to do, from October into November, drive up into the north Georgia mountains, just drive around going nowhere and everywhere and feeling generally good about life. Sometimes we'd get a jar of preserves or mountain honey from a truck farmer by the roadside, or stop into some small local restaurant where the food is bound to be as fresh as the service.

Such a day it was when we first met John Carpenter. It seemed the result of accident, beginning with a low rumbling through the hills.

"You hear that?" I asked Ron.

"Yup," he answered. "Sounds like thunder."

"It's not *supposed* to rain today," I informed him. "That's what the weatherman said."

"I don't guess he said it loud enough," Ron replied. "The clouds didn't hear him."

The thing about the Appalachians – in case you haven't been there yourself – is that you can't always tell what's coming. The hills and hollers hide things from you. Maybe it's started to rain in an adjacent valley and you're sitting in the sun having a picnic and thinking the world is one big ball of sunshine. Then, before you have time to put away the fried chicken and the potato salad, you're underneath a lightning storm, maybe even hail in the spring, and all you can do is pull everything together in the blanket and run for the car and sort it all out at home, unless you want to run the risk of getting barbecued yourself.

As we moved a little higher I could smell the rain and the

acridness of electricity in the atmosphere, odors that intensified when we drove into a bank of cold air, just like cruising into a refrigerated vault.

"What now, navigator?" I asked Ron.

"Onward and upward," he said.

"Are you sure we ought not turn back?"

"Honey, I'm not sure about anything. I'm just here for the ride. You want to go back?"

"Well...." For some reason I didn't, come to think of it. Maybe I enjoyed the thrill of riding into the storm. Maybe I wanted to be naked on a mountaintop in the middle of the cracking lightning and the bouncing booming thunder. Who knows what I wanted? The sky up ahead had turned so dark it almost looked green. The air was colder, the wind was up, and everything smelled of raw nature getting angry. Still, I didn't want to go back down, and when I saw the place through a gap in the trees, I settled on it right away.

"Look, Ron."

"I can't look, babe, this road twists like a snake and the shoulder is a few hundred feet down. You can practically see your own taillights going around these curves."

"There's a little place with a sign up at the top of the ridge – I just saw it."

"What sort of a place?"

"Maybe a stop-and-go-inside sort of place."

"We'll find out," he said.

We pulled up just as the first raindrops – great big round raindrops like marbles of water – hit the windshield. Ron started laughing. "Check out the name of this joint," he said.

I peeked at the sign: *Saucer Inn.*

"That ought to suit you just fine," he said. "Maybe they've hired some illegal *aliens* you could interview."

I told him to hush and gave him a good solid swat on the shoulder. "Ooooh!" he hollered, and clapped his hand against his arm as he fell sidelong out of the car, swinging the door wide open and rolling underneath it. Then I saw him pop up in front of the hood and go limping off across the gravel to the Saucer Inn as though he'd been shot with an elephant gun, leaving me to lean over the front seat into the rain and pull the door to, then get out and follow him.

Ron stopped on the porch and appeared to have recuperated in one of the rockers by the time I reached him. You could really smell the rain now, not just the electric aura of it but the wet holy stuff itself. To tell the truth, it's a scent that clings here all the time, even if only faintly -- that wet, leafy, old-timey smell, as if the hills, having been washed by rain so many centuries, took on its odor always, like a kind of cologne.

The whole northern third of the state smells like a creekbottom. The last time we went out West, I remember, the air was so open it smelled like nothing at all, except whatever was floating through it at the moment, and maybe some distant sea salt. But when we got off the homebound plane and out into the air, even there in Atlanta, you could hear the frogs and smell that creeky water all around and the smoke from some barbecue joint. I think Mars will smell just about the same.

The Saucer Inn was built of black locust wood, rough cut, roofed with slate, and furnished with a wide covered porch made of seasoned two-by-fours, where we were now both seated in our rockers and looking out into what had become an abject downpour. As best as I could recall, there were only two or three other sad-looking cars out there on the gravel between the roadside and the inn, but it was impossible now to see even as far as the road. The wind came howling along the porch, hitting Ron, who was upwind from me, full force. We were right smack on top of the ridge.

In the window behind Ron's head was a red-and-white wooden sign hung by a chain. It said "Open". Just beyond that was a door with a wreath hung on it and a small banner, with the word "Welcome" stitched into it, hung on the wreath. They were open and we were welcome and the scenery had disappeared behind a curtain of water and the cold wind drove the rain up under the eave. Ron propped his boots up on a porchrail.

"Don't get too comfortable," I warned him.

"How's that?" he answered.

"Before you take your nap, put those legs down, I'm going in."

"Still a tad early for supper, isn't it?" he said, still with his legs up. "Besides, I think it might be a trap."

"Well, it's a warm dry trap, babe." I said, standing up and thumping his boots off the rail with a sweep of my arm. He hooted and I heard his heels clumbering along behind me to the door.

"They saw you coming," he was saying. "They sent the storm and disguised the saucer as a restaurant and now they're driving us into it!"

"Hush!" I had started giggling, and the last thing I wanted was to walk into a diner somewhere in the mountains laughing like an idiot. When I start giggling – I mean once I get going at it – there's nothing to be done about it. You just have to let me run.

So I was doing my best to keep myself under control when we stepped into the room, and at that point you couldn't have gotten me to giggle in feathered underwear. It was like stepping across the threshold of another world altogether.

Ron tumbled in behind me, laughing. Then the door banged shut and his laughter in the interior silence that replaced the roar of the rain, was like a pistol-crack, dying away quickly, soaked up into those old wooden walls, which were rough cut like the outside, except the wall opposite the door, which was mostly windows. Ordinarily, I suppose, the little dining room would be pleasant enough, with its old fireplace, and the smell of food, and sunlight outside bouncing off the hills and off the stained-colored-leaves and filtering leisurely through the windows – can you even imagine the view? – to settle in a bright warm patch on the wood of your table, which countless strangers have polished with their sleevecuffs. You think to yourself, *Sometimes in this life just opening your eyes is an indulgence.*

But had we not walked into that brightly lit room with all its little patches of sunlight and its view of the hills and the blue sky away off. Instead, we had wandered into an aquarium inhabited by people. All that could be seen from the windows was water. Almost no light entered the place through the windows now, so that all you saw in them was a dim liquid motion. Two rings of candle-bulbs burned in a pair of wagon wheels hung horizontally from the low rafters. The only other illumination came from between the kitchen doors and from a glass-ensconced candle on each table. Only two of the tables were occupied, one by a young couple with a baby, the other by an older couple dining alone in the corner. I saw the older man move, when a rivulet of light slid swiftly down the window glass and the room flashed like a lightbulb. He was standing up, still wiping his mouth on a napkin, when the concussion – it was more of a physical shock than a sound – the concussion, as I say, hit us. It was as if the thunder had broken out from inside my own body. The air smelled like a battery, and my teeth were set on edge, and then all was as it had been before, except the baby was now howling inconsolably and here was this man standing and watching us.

I had expected him to make some sort of gesture, thinking maybe he owned the place and was having a bite with his wife, business being slow like it was, and that he might perhaps even come over to greet and seat us. But his intent seemed to have been already achieved by the simple act of getting to his feet. He appeared fully satisfied to stand right where he was, napkin in hand, still chewing, and stare at us. It kind of took me away from myself for a moment.

"The sign says *Seat Yourself*," Ron said.

"Well all right, then," I replied. It was time to dive into the aquarium with the other fish. The watery motions from the windows refracted the light in strange shifting patterns around the room. In my mind, I was swimming through a dim liquid air over toward our table by the kitchen, away from the man in the corner.

Light flashed in the room again as we took our seats, but it wasn't lightning. The kitchen doors had swung open momentarily, allowing a silhouette to emerge, which, when the doors swung shut again, took on the fully fleshed shape of a middle-aged woman in slacks and a white blouse, wearing rubber-soled shoes and carrying a pair of menus in one hand.

"Looks like y'all picked the right time to stop," she said as she laid the menus down. "Sugar, let me take that wet coat off your chair for you." Without waiting for an answer, she lifted my jacket from the chairback and carried it across the room to a line of pegs by the front door. The man in corner had reseated himself, but both he and his wife were looking in our direction. The young woman had just about gotten her child calmed down, when another flash and another *boom* set the poor thing off again, howling as wildly the storm itself.

"Ron," I said, "did you see that man stand up when we came in?"

"Uh-huh." He was looking at his menu.

"Don't you think that's odd?" I asked.

"Maybe he's just polite," he suggested. "Isn't it supposed to be good manners to stand when a lady enters the room?"

"I wouldn't trust you to know."

"Hey," Ron said in surprise, dropping the menu and holding his fork up to his nose as if he'd never seen one, "what's this thing here?" He sat staring at the fork in great wonder and curiosity. "Wonder what it's fer," he exclaimed, and made like he might stick it in his ear. I was worried that some of these mountain folks might think he was having fun at their expense. A large man in white pants and shirt had come from the kitchen behind the waitress with a bin of ice and was gathering plates from a bus tub, but he didn't even look up. One thing about Ron, he never lets me get too serious. There's no telling how many bad moods of mine he's ruined over the years, just when they were gaining some momentum. But I forgive him for it.

The waitress was floating noiselessly back toward us in her rubber-bottomed shoes, making the place seem all the more aquarium-like. I half expected her to lift off the ground and sail over to our table like Venus on her clamshell. But with her hair all up in a netted bun, the effect wouldn't have been quite the same, I don't suppose.

"Can I get y'all anything to drink? Some tea?" she asked. We nodded. "Sweet or unsweet?"

"Sweet," we both said in unison.

"Oh, good," she said. "I don't fool with the unsweet tea unless someone asks for it. My name's Pat, let me know if you need anything." She strode efficiently back into the kitchen, catching the door behind the man in white, leaving us there under the glowing wagon wheels.

By now the young family had finished, or had just had enough; she was busy bundling the sniffling baby by the door while he fished some bills out of his wallet for the table, obviously in no hurry to get out since he was the one who would have to fetch the car and pull it around in the downpour. Adjusting his belt, the young man ambled lazily across the dimly-lit room, unknowingly contributing to the illusion of submersion that I'd been nursing. Then the door opened and the sound of the storm poured in along with a rush of cold wind. When the door slammed shut, the noise and the wind slammed shut with it, leaving behind the drone of the rainfall on the roof. We watched their figures disappear into the sheets of rain on the other side of the front window, where the sign, facing us, read "Sorry, Come Again".

"Ron," I whispered, "He's still looking."

"Who?" Ron said distractedly.

"*Who?!* Ron, have you noticed that there's only two other people in here besides us."

"Ah, him," he said. "Maybe he thinks he knows us."

"My stars!" I gasped. "You know what?"

"Don't tell me you know that guy."

"I think I just might." All this time I'd been looking at him sidelong, not wanting to lock eyes with a strange man staring at me from the corner of some subaquarian mountaintop dining room. But now I met his gaze, and although I couldn't place it, I knew I had a memory of this man. Seeing me finally return his glance, the stranger beamed a big joking smile, nodded once, and returned to his meal at last. His lips moved momentarily, whereupon the woman seated at the same table gave us a short look and a smile over her shoulder.

"I wouldn't be surprised if they were family," Ron said, as Miss Pat swished her way back into the room with our glasses. "Wherever you go, coincidences follow you around like a pack of hungry dogs."

The food was good, just your basic home cooking, and we had just about finished our suppers when the wagon wheels went dark and the sliver of light from between the kitchen doors vanished. I had opened my mouth to speak when the *boom* came. It sounded, and felt, as though the mountain had been blown out from beneath us.

"Jay-El!" we heard Pat's voice call in the darkness, and a muffled response that I couldn't make out. I heard the front door open and noticed a figure stepping out into the storm. The woman in the corner sat by herself, seemingly unconcerned. Pat came back into the room and was lighting a couple of hurricane lamps when the older man – it was he who had gone outside – returned.

"I think it hit your transformer, Miss," he said in the waitress's direction.

"It ain't the first time," she said. "Jay-El, you got the gas cut off back there?" she hollered toward the dark kitchen.

"Naw," came the answer. "Gonna leave it on this time. Is the insurance paid up? Fetch me one of them lamps."

"You just stick to burning the food, not the building" she called back to him. "He's a smartass," she said to me.

"I know the type," I said. Miss Pat gave Ron a scrutinizing look while he put on his best *who-me?* face.

"Y'all don't worry," she told us. "I got plenty of lamps and Jay-El can fetch some wood from out back."

"We're about done anyway," I answered. "How're you doing, Ron? You about ready to pay and go?"

"Go?" The older man, who stood looking out the window by the sign, was speaking again, this time to us. "Did I hear you say *go*, little lady?"

As it seemed none of his business what we chose to do,

I was a bit taken aback by the question.

"If you don't mind me saying," he went on, walking toward us, "that road out there is practically a river by now in some places, I'm sure, and I wouldn't drive down that mountain during a storm like this in a car like yours for a million dollars plus change."

"You know something," said Ron, looking up from the table and tilting his chair back. "Alex here has been saying how she thinks she might know you."

"Do tell!" said the stranger.

I could have kicked him. What was he getting us into now?

"That'd be nice," said Pat, "we all might be here for awhile." Then turning to the kitchen, she called out, "Jay-El, you want to dial up Ray and Kitty, see if they got in ok?"

"I'd love to," came the disembodied reply from the darkness, "if I could see to find the phone."

"Sorry," Pat hollered, and went back into the kitchen with a lamp.

"My name's John Carpenter." The stranger held out his hand to me, which I took. His hands were solid, almost square, and his skin was tough, but his fingers encircled my palm like a mother cuddling her baby. Lightning flashed again as I withdrew my hand, and John gave me a quick wink before the thunder hit, so when the *boom* came it seemed more amusing than bothersome. The men shook hands as the thunder rumbled over, under, and around us, and John's wife, whose name was Karen as it turned out, came over to say hello and everyone was introduced all around. Within half a minute I couldn't recall what had spooked me so about John.

"You say you're wife here thinks..." John started to say.

"We're not married," Ron interrupted. "Almost married."

"Uh-huh," John continued. "Well, your girlfriend here," Ron let it slide, but I smiled, "you say she thinks she might know me."

"It just seems like we've met," I said.

"You get to Australia much?" he asked.

"Y'all live in Australia? You don't sound Australian."

"We have a house there," said Karen. "Mostly we live all over the place."

"On the ocean quite a bit," John added.

"Eating here today is a lot like eating in our galley," Karen quipped. I wanted to tell her about my fantasy of the aquariumized room, but decided to keep that to myself for now. Ron's a darling, but he'd make fun. Like most men, it seems he approaches the infinite phenomena of this world either with teeth-gritting seriousness or with a sort of flippant dismissal. Not much in between. Some things – and this is what he fails to understand, but bless his heart, I suppose he can't help it– are serious in a flippant kind of way, not to be brooded over and studied, but not to be dismissed out of hand either. There are sensations in the human heart that are just as vital as they are fleeting.

But cast not thy pearls before swine, as they say. There'd come a time, I knew, when I could choose to tell Karen anything I liked. She would understand there was something vivid to what I was feeling, and she'd be smart enough not to try to pin it down, but just to let it be, to sit with it and see where it was headed. Standing there in that fashionable ankle-length skirt, in that log-walled room flickering with candle- and lamp-lights, you could imagine her as a reserved woman quite easily – all self-possessed and feminine, but ready at a moment's notice to pull an ox from a ditch if need be.

Pat appeared at the kitchen door. "Y'all care for some dessert? On the house. It'll only go bad otherwise."

"You don't have a generator?" John asked with some surprise.

"We do," Pat answered, "but it's old, like us I guess. It'll take care of the freezer and refrigerator, but only just."

"Can't you fit the dessert in the refrigerator?" Ron suggested.

"What for?" said Pat, in mock offense. "You think we serve leftover cobbler around here?" She was a pioneer woman too, Pat was.

The tall figure of Jay-El emerged, filling up the space in the doorway behind her. He rested his hands on her shoulders, kneading them lightly with large hammy fingers, so that she smiled, and sighed once.

"You ever get Ray an' them?" she asked him in a languid sort of voice.

"Line's out," he told her. She frowned a moment. "Guess I'll bring in some wood," he added. He gave Pat's shoulders a squeeze and vanished again into the darkness of the kitchen.

"I'll bring out the cobbler while it's still hot," Pat told us. Then looking at Ron, she added, "I can put yours in the fridge if you'd rather."

"As long as you put me in there after it," he replied. And with that, she was gone again.

"I'm sorry if we stared at you some," John began, "but we were trying to figure out where *we* knew *you* from."

"You don't mean it," said Ron, shaking his head at me like to say, "See there!"

"You look very familiar," Karen admitted.

"That's so funny, 'cause I was thinking the same thing," I told her.

"We're around here a lot," John said. "We have family. In fact, I just inherited a little property a few miles north."

"Sorry about your loss," Ron said.

"Oh, nothing's lost," said John. "We'll all meet up again. In fact," he continued, looking at me now, "maybe it wasn't in this life that we met."

An electric thrill wound its way through my nerves. The way he said it, smiling but with direct eyes, I knew he was only joking at joking. And the way he was looking, and the sound of the rain on the rooftiles, and the strange light reflected in darting wavering bands through the room, all of it conspired to unmask something in me, so that I blurted without thinking, "You look just like Poseidon!"

And he did, too. The spirit and image – the large head with red curly hair and beard streaked with gray, proportionally large almond-shaped eyes, the distinctive nose and sculpted lips, wide chest and shoulders, all of it was there. Put a triton in one of his meaty hands and you could set him down in a fountain like the king of the sea.

John and Karen both laughed. "Do you know how many people tell me I look like Poseidon?" he said.

"I was telling John you look like Aphrodite," said Karen. "I think it's the hair. Our little girl used to have hair like that, but when it darkened some she decided she wanted it cut short...."

Jay-El had come back with an armful of split logs. Ron and John took the chance to escape what had become a woman's conversation by volunteering to help him. They passed Pat on her way into the dining room with four plates of apple cobbler and a gallon tub of ice cream balanced on her arms. "Oh no," Jay-El told the other men, "too late." Our husbands followed him out through the kitchen.

While the men stacked a sizeable pile of wood and prepared the fire, we three women had a fine conversation over some of the best apple cobbler I've put in my mouth. Pat and Jay-El, I discovered, had a house just behind the *Saucer Inn*, which they'd owned and run for the past seventeen years after Jay-El took early retirement from the phone company in Greenville, which they foresaw was on the brink of a development boom, something they'd just as soon skip. So here they were, where overhead was low and their kids and grandkids would actually come visit. As long as the restaurant paid the taxes, they were happy as the proverbial clams.

The way we set it up that evening was that Pat brought us some blankets and pillows and four cots that they kept in the house for when family came. They offered to share their home, but we didn't want to impose – or for that matter to run through the storm – and we still held out hope that the weather might abate and we'd be on our way and in our own beds before too long. Before returning to the house, Pat handed me an old cowbell almost a foot long.

"If anything happens," she said, "you just step out on the backporch and flang that thing around as hard as you can. We'll hear you and come a-running."

Chapter Two
The War

By the evening, of course, we all knew one another fairly well, in the way that you get to know strangers you bivouac with for some short time. Ron and Karen were showing signs of drowsiness, but I for one – maybe it was all the sweet tea – felt not a bit sleepy. Instead, I felt electrified somewhat, perhaps by the storm that still kept up a remarkable fury outside. John looked fresh as a daisy, as if it were ten in the morning rather than ten at night, and it was clear we were likely to be up yakking till all hours. With the others drifting out of the conversation, our talk turned from matters of general interest to topics of our own particular liking.

"I know Ron here's in the music business," John said at one point, "but you never did say what you do."

"Sometimes I work with Ron," I told him, "but mostly I write."

"What about?"

Here was yet another point in time when the universe conspired against me, although to my own good. In my mind, I formulated a straightforward informative response to his inquiry, and my brain – I am quite sure of this – sent it on its way. But what came out of my mouth was, instead, "I write about people like us." Where that phrase emerged from, I couldn't say then and I can't say now. In fact, for half a moment

I thought Karen had said something.

John showed as much reaction as if I'd said I wrote about pottery, or Dutch elm disease, or any other old thing in the world. "People from out of town," he said. Then, suddenly tickled by his own words, he broke into a fit of laughter. I couldn't help but laugh along with him.

"What're we missing?" Ron asked. Jay-El had helped us carry an old couch in from the restaurant office before he'd gone to the house – most likely it had seen more than its share of afternoon naps – and we were all seated on it, under our blankets, John and I in the center and the other two flaking out against the overstuffed arms.

"You from out of town?" John asked Ron, with a short chuckle.

"Depends on what town I'm in at the time," Ron replied.

"Ain't that a fact," came the reply.

"What town *are* we in?" Ron asked.

"We're in the *Saucer Inn*, that's what we're in," I said.

"Uh-oh," said Ron, "we could be just like that cobbler... halfway to Uranus by now."

This time I did punch him, but John nearly doubled over from laughing.

"I think it's time for me to stretch out," said Karen. She kissed John and then stood and said goodnight and glided, still wrapped in her blanket, over toward the fireplace where we'd set up our sleeping arrangements. "Don't keep these young folks up too late with your war stories, John," she added.

"Oh, we haven't even gotten to those yet," he said.

"He'll go all night if you let him," Karen yawned. "Don't

y'all feel rude, just go to sleep whenever you're tired. He'll go for a hike or something."

"You're a vet, John?" said Ron, awake again and slouching forward on the sofa.

"Twenty-one years in the Navy. I was a chief warrant officer and one of the first UDT men."

"UDT?"

"Underwater demolition."

"Tell us a war story John," Ron said, leaning back and reaching his arm around me so I could rest against his body.

"Oh, I ain't really got any war stories."

An unbelieving laugh sailed over from the fireplace.

"Where were you?" Ron asked.

"Pacific," John answered. "I was a boat cogkson third class."

John said, politely. "You heard of Bougainville?"

"Oh yeah," Ron said. "One of the best planned and executed attacks in military history, that's all."

"I'm surprised you'd know that," said John.

"I am too, but I do. You fought there?"

"Not if I could help it," he joked. "Actually, we were there behind the Marines, all of us come over from Guadalcanal, which had already been taken by then, at the end of 'forty-three. They loaded up the gear and rations for those boys, just enough for three to five days for the first wave, and bundled it all into nets that could be loaded quickly into the amphibious landing crafts once we made the Solomons. The boys didn't have any camouflage, so they dyed their undershirts green, painted yellow and brown splotches all over their fatigues, even used vegetable powder to color their skin green. They were a sight. Just about scared the s--- outta *me*, and I was on their side. I'd hate to think what they looked like to the Japanese.

"We knew Bougainville was weak, only about twelve-hundred soldiers at that point, and recon had told us there were no installations of any kind. We didn't quite believe that, and it's a good thing too, because they'd built a string of bunkers out of ironwood and coconut trees, just like they'd done at other places, and manned them with overlapping-fire machinegun nests. But you couldn't hardly see a thing on account of the jungle. What we had to do was deliver these boys up onto the beach, where they'd been trained to hold down the machineguns with firepower while a handful of them crept in close enough to lob grenades into the nests.

"Meanwhile, more men are coming up behind, along with bulldozers--"

"Bulldozers?"

"Yes indeed, and not only them but Amtrac's too – that's amphibious tractors. That island was nothing but a flat spot of jungle. Once they broke the lines, the men were out of sight. So air support, in addition to the mortars, were dropping white-phosphorous flares on the enemy, and colored smoke on our own lines so they could see from the ships and the planes what was happening and who was who. The bulldozers – which would also build the airfield, you see – would tear down the jungle, and at the front of each column of men was a fellow unreeling a spool of telephone wire, and another fellow at the back with his own spool reeling it in again. Commanders would come up and snap their phones on and off the wire, so there was a solid wall of communication. Every once in a while, a man in every unit would send up a small balloon so they could have a clear picture of friendly positions for directing operations.

"Now that might sound all nice and neat, but I can tell you, it looked like hell. Absolute hell. Some LST's down the beach were unloading small tanks, and right behind them some larger tanks with a kind of buckshot in their guns, only it was manshot. Too light to pierce tank armor, but heavy enough to turn a man into a sieve. We knew from experience what the Japanese would do, and sure enough, they swarmed the first fleet of tanks, trying to get in under the guns and drop grenades into the cockpit. Well, they let them get onto the tanks all right, but before they could make the hatches, the tanks in the rear cut loose with the buckshot. You simply can't imagine the slaughter.

"Meanwhile, we're in an old wooden boat that's been resurrected as a landing craft for the War, dumping men out onto that beach under a noise like the end of the world, mortar and bullets and the white-phosphorous and colored smoke, these green men and the machines pouring out into the thick of it. We were damn glad when the last one was off and we could turn tail for the ship."

"You take any hits?" Ron asked.

"We took a few," John said, "but that old Higgins Company boat was a pretty good craft, and once the load was off, as long as you kept her speed up she wouldn't take on water too bad.

"This one mortar nearly took us out, though. It passed wide and blew the stern out of one of the incoming landing craft. Before we knew it, the sea about fifty yards away was alive with Marines. Full packs, guns, gear, their boots were full of water, too far from the ship and too far from land. I swung around and the other fellow in the boat – I don't even remember who he was now – started calling out about the water, that we were taking on water, which we were. And we took on even more as we slowed. There was a film of dissolved vegetable powder on the surface, if you can believe it. I tried to circle at a fast enough clip to get those men aboard and still keep on top of the ocean, but it was close. Too close for my buddy, I think, who looked about panicked by the time we got the last man aboard.

"In fact, he was scared I might try to make the beach again and shouted at me, 'Please don't tell me you're taking them in.' Of course I had no intention of taking them in, but he wasn't too sure at that point."

"I don't blame him," I said. "Id've been scared senseless. You weren't scared?"

"Like a dog shitting peach pits," John replied. "But I couldn't let those men founder. All I could do was lay that throttle out and pray to Almighty God."

"Don't tell me," said Ron, who had his eyes half-closed and his head leaned back against the top of the sofa, "you died."

That made John laugh. "Not just yet," he said, "that's coming up. No, we made it, all of us. Of course I don't imagine all those boys made it home in the end."

"Did you lose any friends in the War, John?" I asked.

"Oh, yeah. Who didn't? I'll never forget this one fellow, a buddy of mine, from Oregon. One evening before another amphibious landing, he hands me a letter, already sealed and addressed to his mother, and asks me to swear I'll get it to her. I'd never seen him like this, and I said, 'What's making you so fatalistic all of a sudden?' He says, 'This one's it for me John.' I tell him, 'What if it's me that gets hit? What if they hit me right through this pocket here and your letter gets too stained with my blood for your poor old mama to read it? Now how do you think that's going to make me feel in the hereafter, knowing I betrayed a trust? You don't want to send me to Jesus with a burden on my soul, do you?'"

"But he wasn't having any of it. He'd made up his mind and that was that."

"Was he right?" Ron asked.

"Oh, he was right. The next morning we found out that this particular beach was better fortified than we'd thought. I was driving a landing craft still, keeping an eye on my buddy from time to time, who was piloting another craft a little ways behind me. He never once looked at me once we were underway. He didn't look scared, or distracted. He was just looking straight ahead, toward the beach.

"I don't know if he saw the shell or not, but it came right at him. Took his head clean off his body. I was watching. His head was gone, and every time his heart pumped, a jet of blood streamed up into the air – I could see it from where I was – but the body didn't seem to notice. It just kept on piloting that boat, straight as an arrow, his hands still on the wheel, driving those Marines on to the landing. Another Navy man managed to shake off the shock somehow, and he took the wheel. When he did, my buddy's body seemed to relax and it slid down onto the deck.

"There's a lot that I've seen, but nothing to equal that. I hope never to see its like again."

"Did you ever talk to his mother?" I asked.

"No, I didn't. I did make sure she got the letter, but at that time, I had no way of talking to her."

"What about the part where you died – didn't you say that was coming up?" said Ron.

"Oh, that was when I was flying. I was shot down and they had me on the operating table for eleven hours. I was in really bad shape – a shattered hip, shrapnel in my kidney and just about everywhere else, a bullet lodged right by my heart. Finally they gave up on me, declared me dead and pulled up the sheet.

"Well, I knew I wasn't dead, or that at least I was aware of something, but I couldn't move or speak or see or hear. Next thing I know, I feel someone lifting me up bodily. I was horrified. 'They're carting me off to the morgue' is what I thought to myself.

"Now, I can't really describe what happened after that, but I'll do my best to translate it into here-and-now terms. I opened my eyes – they weren't these eyes, but they were my eyes nonetheless, and I opened them – and I was travelling behind this being – you could call it a person, a spirit, a soul, and you'd be right and you'd be wrong – I followed it into a grove of... just enormous, ancient oaks. I can't begin to tell you.... They had to be two thousand years old if they were a day, just dripping with moss and resurrection fern. The whole place was awash in light, suffused with it, but not coming from any one place. Everything, it looked like, was made of light, built of it, a solid sort of light. Not reflected, like sunlight, but originating from every vibrating atom, so the world took on a liquidy aspect. My own body looked like this. And my body was whole, not shot half full of metal.

"It took me some amount of time – but time's a problem too – to notice I was hearing music. There's no words for these sounds. As I listened, I found I could never quite catch the melody. It always slipped off, just when I'd think I had it figured out. The song seemed to be writing itself as it was sung. Then I knew, the song I was hearing was made of the light I was seeing. It was growing, really, not singing. I was listening to these atoms of light shifting amongst each other in infinitely intricate but perfectly harmonized patterns.

"I figured then that I was dead. Really dead. And that pained me. The being I was with sensed that pain and asked me about it. I told him – we weren't talking, but I told him – that there must be more for me to do. Then I felt the pangs of pain and fear enter my body again. I knew I had a choice at that moment, to go back to that, or to stay where I was. And I knew, too, that it didn't matter one way or the other what I chose, that God's plan would be accomplished whether I returned to my loved ones or waited for them to come home to me. But it was that – the thought of my family – that pulled me back. I wanted to be with them, so I'd made my choice before I even knew it. All the bodily sensations returned, and I opened my eyes again, my physical eyes now.

"I pulled the covers aside a little bit with my good arm, to get a look at myself. You should have heard that nurse scream. She dropped something or other to the floor, one of the doctors came rushing in, and a couple more nurses."

Ron pulled his head off the couch and looked over at John. "Sounds like you and Alex have a lot in common," he said.

"I wouldn't doubt it," is all the reply John gave.

"Well, folks," Ron said, with nothing in mind to say afterward. I felt him begin to stretch and knew that would be it for him.

Chapter Three
Upbringing

Ron put a few more logs on the fire before wrapping himself up in blankets on his cot. I sat up with John. From time to time John would go out onto the porch, saying he wanted to "breath some lightning", and I guess that's what he did out there.

He came back to the subject of writing. "I've always wanted to write," he said.

"Why haven't you, then?" I asked him.

"Can't sit down long enough."

"What would you write?"

"Why don't I just tell you and you write it?"

I thought about that for a minute. Or really, I thought about it for a split second and then pretended to think for a minute longer. "Why don't you tell me about growing up?" I said. He obliged.

"That could have been a lot shorter story than it turned out," John began. "When I was born, it was in a coldwater shack. It didn't have plumbing or any of that. When it was time for me to be born, they were already scared, because I was an eleven month baby, weighed over thirteen pounds and was twenty-four inches long. I was delivered by an old veterinary doctor. I remember my grandmother saying he had to cut my mother almost to pieces just to get me out. She said it cost my daddy three dollars.

"When that old horsedoctor got me out, I was blue and wasn't breathing. He said, 'This baby is dead,' and he laid me out on the bench there and went back to try and save my mother's life. She'd been in labor seventy-two hours with me.

"So while I was lying there lifeless and cold on that old wooden wet bench, my grandmother went over. She picked me up, looked at me, and held me to her. Then she put her mouth over my nose and my mouth and started breathing. She breathed into me, and pressed her hand against me, until I was breathing on my own.

"My grandmother was born in 1857, before the Union-Confederate War. My mother was her twelfth child. My grandfather's friend Mahappy was there along with my grandmother, that being my mother's mother. My father was of the Cheyenne nation, so he wasn't present at the birth. But I'd always heard about it from my mother and from Mahappy. To tell you the truth I never gave it much thought growing up, until I learned CPR in the military. How she knew what she was doing, back then, I'll never know. She always said angels showed her what to do."

"Were you sick?" I asked.

"No, I was fine as frog hair," he said. "I guess being so long inside my mother, I was supercharged with all that mother-goodness. By nine months I was talking pretty good, my grandmother says. At three years I was reading the Bible. It was my grandfather who taught me to read so young. He was born in 1853. He was Irish, and was taught with a blue-back speller how to read, and that's how he taught me. He'd sit me on his knee and read to me, and he'd talk to me just like I was grown.

"I soaked it all up. Just absorbed it like rain into the ground. One time, I was three years old, a friend of my mother's was at the house. She asked if she could borrow my mother's breast pump. I was so little they weren't paying me any more attention than the man in the moon. But when my mother said, No, she didn't have one, I stopped my playing and I said, 'Yes we do.' They didn't know what to say, a little three year old boy running off to fetch his mama's breast pump. I climbed the ladder up to the loft and dug down to the bottom of things and got the breast pump for the lady. I don't know if I still knew where my mother put it after all that time, or if I knew it some other way.

"Sometimes I'd see what folks were thinking. Sometimes I just remembered. On my first test in school my teacher gave me a zero. I went to her and I said, 'Miss Larks, why did you give me a zero?' She said, 'Because you cheated.' I said, 'No ma'am, I don't cheat and I don't lie either.' She said I had everything just like the book, down to the very word. I didn't know *everyone* didn't remember what they'd seen and read. I told her, 'I don't have anything on me. All I have is this ragged old shirt. You can ask me all the questions again.'

"After eight questions she stopped and said I was a 'gifted child' with a special memory. Then she found a newspaper and had me read a couple of paragraphs and hand it back to her. When she asked me what I'd read, I recited it.

"In my day they said you were a 'gifted child'. Today that phrase means something different, but back then it meant you'd found your gifts from God a little early. If I did, I give the credit to God, and my family, and Mahappy."

"That was your grandfather's friend?"

"My grandfather, yes, my mother's dad. He came from North Carolina in a wagon around 1880. The land boom from the gold rush went bust when they struck at Sutter's Mill in 'forty-nine and everybody ran off to California to go broke instead of going broke at home. Meanwhile, the Army was trying to get farmers into the hills to keep the poor Indians from coming back to where their people had lived forever, God bless their souls. Even gave the farmers dynamite to blow stumps out of the ground so they could clear the land for tobacco and houses and family farms. Once it was clear that the white man had won the day, every dirt farmer from every little sorry used-up patch of red sharecropper's clay from three hundred miles around and rode up into the hills with every pin and button he owned piled onto a wagon, sometimes only a cart, and the whole family piled on top of that to keep the wind from taking it.

"Some of those wagon-trails got to be pretty well used, and people had businesses along the way if they owned some land there, liveries and little general stores or groceries. There were waterholes, even regular church meetings at certain spots along the road, where maybe there was only a cross made from granite flakes to mark the church. A lot of folks never made it up into the hills; they'd settle somewhere along the way, or turn and follow a river.

"One time my grandfather was stopped at a cooper's to see if they could help him fix a cracked wheelband, and while they were working on it, he got to talking with the little fellow who worked there who was helping him. He wasn't but just over four foot tall, this man, weighed no more than a hundred pounds. But he threw those wheels around and handled that iron like it was putty.

"My grandfather figured he must have been different from birth, the way some babies are. As short as he was, he was a lean little man, even his ears were lean, which made his head seem over-large. And then too, he sported a huge handlebar moustache that covered his mouth, and nobody ever saw him without his cap (we called them slouch-hats). My father always thought Mahappy was bald-headed, that he wore the cap and moustache to compensate. My mother was of the opinion that he wanted to protect his skin from the sun – it stayed an ashy kind of pale year-round.

"Anyway, my granddaddy watched how that little fellow went to work on that wheel, and he says to him, 'You mind me asking how much they pay you here?' 'Nope,' he says, ''cause they don't pay me nothing. I get a pallet and a garden patch.'

"Well, my grandfather was thinking about building a house and a barn and breaking the soil, and Mahappy takes that wheel, rolls it over to the wagon, lifts the axle like it was a pump handle and slides the hub on by himself. 'This is a useful man,' my granddaddy thinks.

"He says to Mahappy, 'We're going on. You can come with us if you want to.' And he did.

"My grandmother says he helped them start the farm and make it through the first few winters. He'd fixed himself a place in the barn to stay in when they built it. And then one spring he took off. Came back right before my mother was born. He nursed her through the scarlet fever – she had it her first year – stayed a few more years and disappeared again."

"You said he was there when you were born," I reminded him.

"He came back for that too," John said. "Stayed till I was eight years old. When my mother got married they built another house there on the land, because my father had no land of his own. It was hard times then. Nobody up there had money. Whatever cash crops they had, went for thread and veterinary bills and such things. Not that they needed much money, when the grist mill took payment in grain, and the store was just as likely to take eggs as coins. But my folks had young ones and old ones to care for and the weather wasn't always kind.

"Some winters I remember people were boiling bark and seed-husks for coffee, digging roots, some killed their horses, even their dogs, God help them. We never got that bad off, even when we were all living in one room and drinking snow-melt. If we couldn't scare up any food close to home, or couldn't get out for the snow, Mahappy would take off into the woods a couple of days and come back with something – a rabbit, some nuts, berries, maybe even a goose. One time he came back with a deer carcass across his shoulders. We butchered it and stowed it in the half-cellar, what we didn't eat right then. Mahappy hadn't eaten a bit of it. How he carried it through those hills I can't say, because he only talked about where he was at, not where he'd been.

"I think I was closer to him than to my own daddy and granddaddy. I remember lying out on the ground, both of us with our hands behind our heads, one leg propped on the other, and him telling me all about the stars. He would name them and tell me about the different planets and constellations. This was the 1920's. I knew how to tell my way by the stars, any season of the year, by the time I was four years old.

"Just about that time too, he showed me what he called 'kahth work'. It was like Aikido, but of course there was no Aikido back then. He showed me how to plot a course, which was geometry and trigonometry, although I didn't think of it that way. He even showed me some calculus before he left. And here I was, not but eight years old.

"I didn't think about it like it was schoolwork. That's not what I think about when I remember the man."

"What do you remember, John?"

"I remember –" he was silent a moment, "—one morning, I was five years old, Mahappy wakes me up before daylight, says, 'Son, get up.' I jump up and put on my overalls. I didn't have any shoes so I was barefooted. He took me a long ways into the woods but it didn't take us long to get there. He had a beautiful black mare in a pasture. The sun wasn't even over the trees, and she'd just dropped a little pinto colt. He says, 'Son, take off your shirt and wipe that little baby colt down.' I looked at him a minute, took my old ragged shirt off and went to the little colt and wiped it down. I was trembling. The mare wasn't even nervous about me.

"Then Mahappy says, 'Do you like that baby?' I stuttered and said, 'Yes sir, I do.' And he replied, 'It's yours.'"

John stopped. I was seeing him as a little boy standing in that field, barefoot, shirtless, with his coverall straps hanging loose, the sunlight just catching the colt as it wobbles up. He must have been the picture of wonderment.

"Tell me something, John – when did you last see Mahappy?"

"I was sixteen, going into the military, and he showed up again, even though we'd moved a lot since then. I'd finished school early, the year before. He only stayed a couple of weeks and then he was gone."

"You think you'll see him again?"

"I don't know. I hope so. I think I will."

Chapter Four
Old Navigators

"I wish it were clear tonight," I was telling John, "I'd let you show me the constellations." I'd gone out on the porch with him to see what breathing lightning was like. I knew what he meant once I got out there. (I highly recommend it, by the way.)

"If it was clear," he said, "we wouldn't be here."

"What do you do, on the ocean," I asked, "if you get into something like this?"

"Go by your instruments," he said.

"What about before they had instruments?"

"Depends on who's sailing, and when, and where, and in what."

"I'd hate to be in a boat," I said. "In fact, I want to be inside."

We went in and John stoked the fire, and we sat at opposite ends of the sofa talking about sailing – about Egyptian reed boats, and the skin boats of the North American red ochre people, the bark canoes to the south, Viking timber boats that sailed to America before Columbus, Charon's raft across the River Styx, Chinese steamboats of the sixteen-hundreds.

After getting things squared away in America, John and Karen were sailing for New Zealand's Cook Islands to visit an elderly Polynesian navigator who still sailed the old way. He was a gentleman in his eighties, one of the two oldest men still trained in the ancient sailing techniques of the Polynesian Islanders. These were the same people who settled thousands of square miles of ocean on mere rafts over 1,500 years ago. Now, it was imperative to find students for the art before the knowledge was gone from the world altogether. And with modern life being what it is, young people having families and jobs and rent to pay and whatnot, it's not easy finding folks who are in a position to learn.

John first met the old man – who wasn't yet an old man at the time, however, but an old-man-seedling – during his Navy years. He (John) was on a 317-foot destroyer about 500 miles out when someone spotted a small craft, presumably a life-raft. They circled for it, and it was this fellow in his sailboat, as it turns out. They sent a boat to fetch him and the crew were half-stupefied to find him in perfect health and spirits. He spoke English, was perfectly familiar with their type of ship, and accepted an invitation to come aboard. All he had with him there was some fresh water, coconuts, fruit, oars, fishing tackle, some equipment for repairing the boat and sails, and a small pile of sand which he said was for cooking fish.

"I asked him if he knew where he was," John told me, "that he was a long way out in the ocean. He said, 'I know.' So I went and found a chart for him to look at and asked him to show me. He pointed to a spot and looked back at me. He knew right where we were. I asked him how did he know."

Their new acquaintance – who, being an islander, was also familiar with the modern navigational equipment of the time, from the ships that went in and out of port – talked with some of the crew about the old style of Polynesian sailing. They sailed by the stars as all sailors do, but they also knew the flight habits of innumerable species of birds and used these to navigate among the islands. In fact, this is how his people had discovered and settled the islands in the first place, shortly after the time of Christ, by reckoning their courses to new land according to the birds and the tides. A trained sailor, back then and today (tomorrow, we shall see), would read the patterns of the waves in the big middle of the boundless ocean, the way an American sailor reads a radar screen. Then, he would immerse himself up to his neck in the sea, to feel the subtler undercurrents with his body.

All of this, the American sailors could understand. They could not have done it themselves, but they knew intuitively that it was possible with the proper training. What they could not understand was the business of the stones. He told them, John said, that "he would cast stones and each stone would represent a certain star. He did this on the ground, before ever taking off. As the stars came up, he would navigate a course by the chart of stones on the ground. Then he would memorize the chart. Then leave."

He was a native of the Cook Islands, and he told John where he might find him, which John did a few months later. John wanted to learn to navigate Polynesian style. And when he had the time, many years afterward, he went back and began.

And it was from this man that John learned the story of the Mystery of Easter Island, which is not the same for Polynesians as it is for Europeans. When Roggeveen's

Dutchmen arrived that Easter of 1722, the Mystery was already beyond the memory of the natives. There were snatches of songs in the old language, but no one knew what they meant. They sang them as lullabies, or to children's games. One game, a sort of cat's cradle, was supposed to have been handed down from the old religion.

There were stones with ancient writing, but no one alive knew how to read the hieroglyphs – or anything else for that matter; they were a completely illiterate people. And yet, obviously, here had been a great civilization, capable of carving monuments as large as fifty tons out of lavastone quarries with stone tools and transporting them for miles to the seashore. And not only that, but capable also of capping these monuments with stone "topknots" weighing several thousand pounds.

What a unique feeling it must have been, for old Roggeveen, stepping out onto that unlettered pagan island – and not only pagan, but pagans who had abandoned even their own pagan religion – on the day of Easter Sunday, over a century before Charles Darwin ever published a book, and holding service with his crew right there on the sand. And looking up and knowing, instantly, that these people, whoever they were, had as much chance as he himself of knowing the craft behind the awesome architecture.

And what an odd sight for the people living on that part of the island. Imagine it. Up sails the biggest boat you've ever seen in your life, manned by a group of fearfully pale men wearing outlandish clothing, who speak gibberish and hold a ceremony by the seaside, the focal point of which is the image of a man nailed in death's agony upon a cross. They smell horribly. They would like some fresh fruit. The fresh fruit gives them diarrhea, so they have to be taken care of. You have no idea where they're from, or why they're out here, but they have guns, which you've never seen (or even imagined), and one of them has shot the young widow's best sow, and cooked it and shared it with several of the other men – without even offering the poor woman the bones, which they threw to the dogs, laughing and happy. No doubt, you are glad when they go away, but the world seems more frightening than you had ever thought it might be.

But once these two peoples learned to speak to one another, it was clear that no one on Earth had any idea about the statues, except that they were inhabited by certain spirits, and their power was concentrated in their eyes. Most of the eyes had been put out, and these monuments were considered "dead". The few "living" ones on the island housed ancient spirits of great power, to whom offerings were made of flowers and prepared food. Statues with one eye, or a damaged eye, were the abodes of evil spirits, demons of fever and blindness. During certain dark epochs, the islanders said, the sick were abandoned on the hillsides at the feet of these cracked idols as a propitiation to the Devil. In that way, they had survived many plagues.

The Europeans, as it turned out, would introduce the already beleaguered islanders to a few more plagues over the following two and a half centuries. In return, the Europeans would eventually help solve part of the Mystery, which was, *What was the old Religion? What was the old Language? Who were the old People?* European (as well as African, Asian, and American) archaeologists and linguists helped uncover the clues and decipher the signs. Women and men like the navigator attached meanings to these signs. In that way, the past was uncovered, like the sand washing off a buried ship's chest on the shore. And the history they discovered, briefly, was this:

Some time in the first few centuries after the death of Jesus of Nazareth, island-dwellers from the east, using the same sort of techniques that our sailor from the Cook Islands used, discovered what we call Easter Island and settled it as a colony. The island prospered – so much so, that it developed its own culture, its own religion, and eventually (by winning a war with the mother-country) its own government.

What became the unique religion of the Easter Islanders was originally a sect of the vulcan-cults practiced by the eastern tribes of the Pacific Rim, the most massive active volcanic formation on Earth. The settlers immediately recognized the island as an extinct volcano, and the figures quarried out of the lavastone (which, according to their religion, was the divine semen of the creator-god) each held a spark of the spirit of the almighty.

Using stone tools, they quarried their idols and rolled them over treetrunks, by pulling on ropes of palm fiber, down to the seashore, where they were finished and hoisted up, staring out to sea, focusing the potent vision of the island's deity on anyone who approached. How they got the topknots on has never been discovered. It is a feat impossible to reproduce with primitive technology.

But prosperity caused unexpected, serious problems. It was about the turn of the last millenium. *The Pillow Book of Sei Shonagon* was being written in Japan. Rajaraja the Great invaded Ceylon. In India, Sridhara discovered zero. Holy Roman Emperor Otto III, king of the Franks, received a mosquito bite in his campaign against, ironically enough, the Romans; somewhere in that mosquito bite was the malaria that killed him. In Europe, steel plows, some even with wheels, were replacing wooden ones (fortunately for the Europeans, who were running out of wood – or at least for the Europeans who could afford a steel plow). Leif Ericsson sailed as far as Canada, and returned to find his Greenland converting to Christianity under a Norwegian campaign funded by England. Leif's descendants would fail for six centuries to establish a colony in the west.

On Easter Island, the lean cows were eating up the fat cows, so to speak. Just like Europe had done, they had used up their wood. Traders made boats from it. Potters fired kilns with it. People built houses of it. Quarrymen rolled their massive god's-heads to the sea upon it.

The forests disappeared. Coconut fiber, coconut kernel and milk, and coconut oil became scarce. Farmland, overworked for too many years, began to fail. There was hunger. At first, quarrying continued furiously, then abruptly ceased. The population declined, and villages withdrew from one another. Whatever power had enabled them to raise those topknots, by now it had obviously abandoned them. Statues carved and raised during this period had no stones balanced on their heads.

Electing their own chiefs, the different villages sent out men to attack their neighbors' gods, to put out their eyes, and to topple them if they could. This was when most of the angelic spirits were driven out, or replaced by demons. So the old religion crumbled with its idols, and was replaced by the warrior cult of the Birdmen.

Every year, young servants of the competing island warrior-kings swam through the treacherous strait to the adjacent rock islands, the nesting grounds of certain Pacific sea eagles. The first to return to shore with an intact egg, that young man won for his master the Warlordship. His Warlordship's first act was always to sacrifice the victorious young swimmer to the Bird God. His second act, invariably, was the summary execution of all other contestants and of any surviving servants. The losing servants almost always chose to swim out to sea and be drowned or dashed against the rocks. One particular boy became famous by climbing the rock ledges and allowing himself to perish there, becoming too weak after three days of thirst to fend off the attacks of the sea eagles, who consumed him partially while he was yet alive.

Having degenerated into war-camps, the villages became deserted. There was no government of any sort, not even military. A handful of people lived in small cavelike dens in the lava rock, surviving mostly on grass, insects, reptiles, birds' eggs, and at times even descending into cannibalism. The language, culture, history, and religion were forgotten. They were possibly the most forlorn generation of the Earth.

"Suppose, one day, you can't get a compass, you can't get a radio," the old navigator once said to John. "Well, then."

Chapter Five
Travels in the Triangle

I knew we'd be up all night. We had to be if I was going to write a book, and if John was going to have the chance to get one written without having to sit still any longer than suited him, which wasn't very long. He was a vivacious man, up and down constantly, frequently heading out to the porch "to recharge" as he said. I don't think I would have been too surprised to find him out there catching strings of lightning by the tail and draining off their electric juices.

As for me, I caught a few catnaps on the couch while he went on his jaunts. Whenever the wind blew in I was awake and surprisingly clear-headed in the few seconds it took the door to swing shut. I think some of the lightning in John's aura had its energizing effect on me. I'd already figured he must be several years older than he appeared, and I assumed the same was true for Karen. She seemed pumped full of life, even asleep there on her cot. Her face reminded me, in the rare atmosphere of firelight and storm, of one of those mountain lakes, all serene, yet a raven haired beauty like on the Harlequin Romance Novels.

After our last conversation, John had gone out for a while and then returned to sneak into the kitchen with one of the hurricane lamps to fix himself another glass of tea. "I love this stuff," he whispered, smiling, as he re-entered the dining room, crossed over to the fireplace, silently cleaned the ashes beneath the fire irons into the scuttle (don't ask me how), added a couple more logs, and sat for a moment looking at his wife, like a young man seeing his lover asleep for the first time.

"She's a trooper," he commented as he took his place again on the sofa. "Lord knows why she puts up with me. But me and her, we've been all over this world together. Crossed the Devil's Triangle I don't know how many times now with nary a scratch. She don't flinch."

"What happens out there?" I asked. I'm not a sailor, and most of this was new to me, I'll confess, although I had heard stories of unexplained vanishings in the Atlantic.

"It's hard to say," John answered. "Most everyone who's disappeared out there never sent a distress call and left no wreckage to be found." He was silent for a moment.

"What do you think happens, John?"

He said nothing, but after half a minute raised his frame off the couch and strode to the door, removing a napkin from a table as he passed. Cold wind and the roar of the rain swept in as he pushed open the door with one hand, holding the napkin loose in the breeze by fingers of the other hand there in the draft. The sleepers by the fire stirred, but did not awaken. The napkin wavered in the wind beneath John's rockish hand, snapping lightly back and forth as the wind struggled in the dark threshold. Then it was gone, snatched out into the storm by a sudden backdraft. John closed the door again and settled back onto the sofa.

"If you stand in a doorway, you're likely to get caught in a draft sooner or later," he said. "Most folks make it through without incident, but if you're in the wrong place at the wrong time, you can get swallowed. It's that simple. And no one will ever find you."

"But what doorway?" I asked. "What draft?"

"Well now, that's the question, isn't it?" For the first time, he seemed reluctant to talk, so I let him sit with his thoughts a while. Eventually, he said, "You have to understand that it began during the War."

"World War Two."

"Yes. The first large-scale disappearance came just after the War, on December fifth, 1945."

On that day in December, at ten minutes past two o'clock in the afternoon, Flight 19 – consisting of five TBM Avenger torpedo bombers manned by an instructor, four Naval Aviators, and nine enlisted airmen – left the US Naval Air Station at Ft. Lauderdale, Florida on a standard advanced navigation training hop ("Navigation Problem No. 1") involving low-level bombing practice over Chicken Shoals. All men and airplanes, along with the crew and craft of a search plane sent to find them, would vanish without leaving a trace.

Flight 19's original departure time was 1:45 that afternoon. However, the instructor, who arrived later than scheduled (only half an hour before the flight), asked to be replaced, giving no reason for his request, which was denied on the grounds that no other instructors were available.

The weather over the area was recorded as "favorable". The Avengers took to the air at 2:10, and by 3:00 had completed their bombing and were continuing on course.

Shortly thereafter, the senior flight instructor at Ft. Lauderdale intercepted a radio transmission on 4805 kc. A voice was asking Powers (one of the students on Flight 19) repeatedly for compass bearings.

"I don't know where we are," came the answer. "We must have got lost after that last turn."

The instructor at the Air Station (who will be identified by his call sign, FT-74) radioed Ft. Lauderdale that a boat or some planes were lost over the Atlantic. Then he attempted to contact the lost pilots.

"This is FT-74, plane or boat calling 'Powers' please identify yourself so someone can help you."

There was no response to this call, but the first voice came back on the air, asking if anyone had any "suggestions".

FT-74 repeated his message, and this time was answered by FT-28, the instructor of Flight 19. The senior instructor on the ground radioed again.

"FT-28, this is FT-74, what is your trouble?"

FT-28: "Both my compasses are out and I am trying to find Fort Lauderdale, Florida. I am over land but it's broken. I am sure I'm in the Keys but I don't know how far down and I don't know how to get to Fort Lauderdale."

FT-74: "What is your present altitude. I will fly south and meet you."

FT-28: "I know where I am now. I'm at 2300 feet. Don't come after me."

FT-74: "Roger, you're at 2300. I'm coming to meet you anyhow."

Several minutes later, FT-28 transmitted again: "We have just passed over a small island. We have no other land in sight." Since FT-28 had reported being over the keys only minutes earlier, the news that they were now out of sight of land was disturbing to say the least.

After a few minutes more, came this message: "Can you have Miami or someone turn on their radar gear and pick us up? We don't seem to be getting far...."

FT-74: "You can't expect to get here in ten minutes. You have a 30- to 35-knot head or crosswind. Turn on your emergency IFF gear."

c. 4:00 PM, Ft. Lauderdale receives a transmission from FT-28, confirming that the Avengers are lost.

4:26 PM, Air-Sea Rescue Unit Four (ASRTU-4), Ft. Everglades picks up FT-28: "I am at angels 3.5; have on emergency IFF. Does anyone in the area have a radar screen that could pick us up?"

Some 20 land facilities are put on alert. Merchant ships in the area are notified and Coast Guard vessels prepared to put to sea. Still, no one can get a fix on the Avengers.

4:28 PM, ASRTU-4 contacts Flight 19, recommending that a plane with an operational compass take the lead. ASRTU-4 received confirmation of the message from FT-28. However, from the garbled transmissions that followed, it was clear that the aircraft did not alter formation. What is unclear is whether any of the planes had working compasses at this time; since no other plane was designated to lead, it is likely that they did not. Furthermore, as FT-28 repeatedly asked the other airmen to report the durations of their course headings, it appears that the instructor's clock was also not functioning. And while magnetic anomalies could explain a temporary compass malfunction, the extended failure of *both* of FT-28's compasses (and perhaps those of all five Avengers) as well as the failure of the instructor's timepiece (which would not have been affected by disturbances in the Earth's magnetic field) seemed to defy explanation.

About this time, FT-74 was losing contact with Flight 19: "Your transmissions are fading. Something is wrong. What is your altitude?"

FT-28: "I'm at 4500 feet."

After receiving this last transmission, FT-74's transmitter failed.

4:31 PM, FT-28 received by ASRTU-4: "One of the planes in the flight thinks if we went 270 degrees we could hit land."

4:39 PM, ASRTU-4 asks FT-28 if he is equipped with the standard YG homing transmitter that would allow him to determine his location from the tower's direction-finder. No reply is received. Six minutes later, FT-28 is heard attempting to determine his position. He is no longer sure whether they have passed to the east or west of the keys, into the Atlantic Ocean or into the Gulf of Mexico. FT-74, the Naval Air Station, and ASRTU-4 are certain that the planes never left the Atlantic. Still, the Avengers have not been pinpointed, and no IFF can be detected, despite FT-28's report that his is on.

Flight 19 is requested to broadcast continuously on 4805 kc. No reply from the planes is received. A subsequent transmission requests them to begin transmitting on 3000 kc, to which FT-28 replies: "I cannot switch frequencies. I must keep my planes intact."

4:56 PM, FT-28 is requested to turn on his YG receiver. There is no response.

5:03 PM, FT-28: "Change course to 090 degrees for 10 minutes." Two students are heard recommending a westerly course.

5:16 PM, FT-28 instructs his pilots to continue flying 270 degrees "until we hit the beach or run out of gas." Still, no one on land or sea can locate the planes.

5:36 PM, having returned to the Ft. Lauderdale Naval Air Station, FT-74 requests permission to search with the duty aircraft, a one-pilot single-engine ready plane. Both the duty officer and the flight officer refuse permission. Although the lost training flight still eludes radar, it is calculated that the aircraft must be close enough to hit land on their current course before running out of fuel in approximately one hour.

6:04 PM, transmissions overheard from FT-28 indicate another change in course. By this time, radio triangulation has determined that the planes must be in a 30,000 square-mile area east of New Smyrna off Florida's Atlantic coast. All local facilities ignite beacons, searchlights, and ground lights. Search planes are dispatched.

6:20 PM, fragmentary messages received from Flight 19: "All planes close up tight.... We'll have to ditch unless landfall.... When the first plane drops below ten gallons, we all go down together."

Around this time, a pair of PBM-5 search planes are recruited to join the search. The flight mechanic who serviced and fueled PBM-5 BuNo 59225 would testify, "I found it to be A-1. I spent about an hour in the aircraft... and there was no indication of any gas fumes. There was no discrepancy in any of the equipment, and when we started up the engines they operated normally."

7:27 PM, PBM-5 BuNo 59225 takes off from Banana River, en route to New Smyrna with a crew of 13.

7:30 PM, PBM-5 BuNo 59225 radios "out".

No signal or trace of the PBM-5, or of Flight 19, was ever discovered after this transmission. The ensuing five-day search uncovered the wreckage of many other ships and planes, but not a single item from the Avengers or their would-be rescuer.

Although Flight 19 was the first massive disappearance in the Triangle since the invention of modern communications and tracking equipment (there were others previously, but nothing on such a grand scale), it was certainly not the last. And not only planes, but entire ships have vanished from radar screens and radio receivers. The most famous of these ships was probably the *SS Marine Sulphur Queen*, last heard from in the area of Key West, Florida on the 3rd of February, 1963 as she headed for Norfolk, Virginia from Beaumont, Texas with a load of molten sulfur and a crew of 39 men.

The only memento ever recovered from the ship was a single life jacket, discovered floating in a calm Atlantic by a Coast Guard search vessel on February 6th, some 40 miles southwest of the ship's last known position. No distress signals and no sign of fire or disaster were ever reported by ships or planes in the area, or by facilities on land, during the time of the disappearance. Oddly enough, the circumstances of the *Marine Sulphur Queen* are remarkably similar to those of the Navy's *Southern District*, which vanished off the North Carolina coast in 1954, also carrying a load of sulfur.

"Some of the old dogs say the Devil reached up and dragged the two ships down to Hell, that he was running low on sulfur," John commented. "Well, maybe someone needs the sulfur, but I don't think it's Old Scratch."

We talked of these events, and of many other strange episodes at sea. Of various disappearances and strange sightings. Of the Curry party – forced to abandon the British *Lara* in three lifeboats in 1881 after a fire destroyed the sailing ship some 1,500 miles off the coast of Mexico – who survived twenty-three days in an unexplained upwelling of fresh water.

But there was something John was holding to himself now. All of these avenues of conversation, I was sure, were leading to Rome, so to speak. But before catching sight of the domes of the holy city, there was always a detour onto another road. By now I was getting to know the countryside quite well; but what stood at the center? I felt like a haji circling the Kaaba in Mecca – maybe it was something that could only be comprehended by seeing it from all sides.

"What do you make of it," I asked him outright, "all of these mysteries?"

"There's mysteries and then there's mysteries," he replied. "How's this for a mystery at sea? Back when I was in the merchant marine, I knew this one fellow who kept on the good side of the Captain by bringing him his coffee every morning. The Captain liked it hot and black and full up to the rim. Every morning, this fellow, he'd fetch it up to him. No matter the seas, no matter how the boat was rolling, the Captain had his full mug of coffee before dawn. Whenever anyone else brought it up, the liquid stood at least an inch below the lip by the time it got to him. So our friend here was put on permanent coffee detail, and took things pretty easy after that, we noticed.

"So finally we decide we're gonna get to the bottom of this, and we spread out to keep an eye on our man along his morning route. Now, this one buddy of mine was stashed at the foot of the ladder leading to the deck, and I had stowed myself up top. Well, he sees this guy fetch the coffee from the galley and bring it out to the ladder. When he thinks no one's watching, he takes a hearty swig of the Captain's coffee and climbs up on deck. When he gets topside, *sploosh!*, it all goes back into the cup from whence it came, and our fine friend saunters off as cool as you please to deliver it.

"Of course we felt like fools, grown men sneaking around a ship, playing hide-and-seek, trying to get to the bottom of a trick so shallow any six-year-old could have seen through it. But we didn't, you see, because we wouldn't have done it, and we couldn't make ourselves think like a man who would.

"The Captain too, he must've been fooling himself. As long as *he'd* been at sea, he knew damn well it was impossible to bring a full cup of coffee up that ladder with nothing on you but your shirt pockets. And where else was a man gonna stow the coffee but his own mouth? But the Captain didn't want to know. His mind wouldn't let him think of it. 'Enjoy your coffee,' it told him. 'Don't worry where it comes from.'

"And that's your basic elements for a mystery. Someone

withholds information that most people would rather not know in the first place, and everyone's content to let it stand at that. Stephen Crane wrote a poem near the end of the last century, called 'The Wayfarer':

The wayfarer,
Perceiving the pathway to truth,
Was struck with astonishment.
It was thickly grown with weeds.
'Ha,' he said,
'I see that none has passed here
In a long time.'
Later he saw that each weed was a singular knife.
'Well,' he mumbled at last,
'Doubtless there are other roads.'

And it's as true today as it was a hundred years ago."

With that, John got up to tend to the fire. He had sidestepped the question.

"What did you mean," I asked him before he could even sit down again, "when you said, 'It began during the War'?"

"You don't give up, do you?"

"No."

Chapter Six

Out-of-Towners

"When I was in the Solomons, I met a young sergeant, First Marine Division, named Stephen J. Brickner. In August of 'forty-two, he was on Tulagi, that's west of Guadalcanal. This is five years before the Mount Ranier sighting, which was when folks started talking about UFO's because of the way Ken Arnold called them that – he was the astronomer who recorded it.

"The morning of the 10th, Brickner's leaning out of his foxhole, cleaning his rifle. It's one of those crystal clear days like they get in the tropics. Out of nowhere, the air raid siren goes off. No condition red, just the siren. So he slides down in his hole, with his back to the dirt, keeping his eyes out for the formation.

"You could tell the Japanese planes by the engine noise. We used to say they sounded like sewing machines. But he hears a 'mighty roar' – that's his words – and says a few seconds later, he spots the formation. And they're not Japanese.

"Someone calls out 'What the hell is that?' and a fellow in another hole shouts that it's Japanese planes looking for the fleet. The one thing they *weren't* was bombers. First of all, they were too high, way too high. Secondly, they were too fast. And thirdly, it wasn't a couple dozen of them in a little V, like geese, which is how the bombers would come in; it was well over a hundred aircraft, flying in straight-line formation, ten or twelve to a line. He said he couldn't see any wings or tails, but it was only his fifth day in combat, and the objects were at a high altitude, and they emitted a powerful glare.

"There were so many people seeing these things in 'forty-two that they came up with a name for them, started calling them 'foo fighters'. Three months after Brickner, a sergeant on an anti-sub patrol in the Bay of Biscay got photographs with a K-20 camera of a 'massive' wingless object that the tail gunner had spotted. They kept visual contact for fifteen minutes while it gained altitude, and then it reversed direction and took off and was gone. The sergeant said he got a perfect print from an exposure taken with a filter, but American and British intelligence won't declassify the film. So that's that. Enjoy your coffee, Captain.

" A month after that, an RAF pilot – which is to say a British pilot – named Lumsden spotted two of them in a Hurricane interceptor over France. It was eight o'clock at night, he was at seven thousand feet over the mouth of the Somme River. He sees what he thinks is some tracer flak coming up from the ground behind him. He turns the plane, and now the lights are larger and brighter. They're still gaining altitude, and maintaining their relative positions, one slightly above the other. At seven thousand they level off with him. Lumsden returns to his original course, and the lights follow.

"He nose-dives. They stick with him for a couple of thousand feet. Then they actually overtake him. He levels out at about four thousand, they're at about three thousand or so. They come up and level off behind him again. He's going about

260 miles per hour at this point. After a while, the lights fall behind. Lumsden reports the foos, and the next night a squadron flight commander gets chased by a green light.

"It was also in 'forty-two, I don't know what month, when Dirk Wylie, another sergeant, spotted one off the southern US coast. He said it looked like a flare, but flying level with the horizon. He had no way to estimate its size or distance or speed. If it was a plane, he said it would have been moderately fast, at that point. But within a minute of first contact, the light turned ninety degrees straight down into the water. A couple of minutes later, it re-emerged, and ascended in 'an absolute vertical line until it was out of sight'.

"Then in 'forty-four, in April, there was the supersonic scare. Radar plotters at the Argus 16 Combat Intelligence Center at Tarawa clocked a bogey at 700 miles an hour. Everyone caught it, traced it, all the sets checked out, no mechanical or electrical malfunction. At that point, the Allies believed that the Japanese were testing supersonic planes. It wasn't until after the war ended that they knew for sure that no such plane existed. But *something* had gone 700 miles an hour past Tarawa.

"In August that same year, August 10th, a B-29 had close contact in Sumatra. He said in his report, 'my copilot reported a strange object pacing us about 500 yards off the starboard wing. At that distance it appeared as a spherical object, probably five or six feet in diameter, of a very bright or intense red or orange in color. It seemed to have a halo effect. My gunner reported it coming in from about the five o'clock position at our level. It seemed to throb or vibrate constantly.

Assuming it was some kind of radio-controlled object sent to pace us, I went into evasive action, changing direction constantly, as much as 90 degrees and altitude about 2,000 feet. It followed our every maneuver for about eight minutes, always holding a position about 500 yards out and about two o'clock in relation to the plane. When it left, it made an abrupt 90 degree turn; accelerating rapidly, it disappeared into the overcast.'

"You can't even count the number of sightings from there on out. Fliers over Germany, on both sides, got used to having what they called 'crystal balls' join their formations and peel away again. It was happening all over the world in 'forty-four and ever since."

"What does that have to do with the Bermuda Triangle?" I asked. By now, I didn't know if we were getting closer to the center, or if we had deviated onto yet another sideroad.

John seemed animated by the question. I noticed him begin to speak a couple of times, but he was like a man trying to move a thing too big for him to hold. Not something heavy, but too large and sprawling – like one man raising an enormous tent by himself. He was saying to me, *just hold onto that pole right there for me a little while*. I wanted it all up, I wanted the whole tent.

"That war," he said at last, "got someone's attention. Do you see?"

"You mean the out-of-towners."

"That's exactly who I mean. It's happened before. The pyramids at Giza. The airstrips and aerial monuments at Nazca, in Peru. Records of the Nommos from the Dogon people. Ancient hieroglyphs of flying machines, and ancient texts recording commerce with advanced airborne civilizations. All over the world, in cultures that had absolutely no contact with one another, we find the same evidence -- even painted on the rocks by North American tribes who had no written language.

"People through history have been instructed in science, architecture, and religion by outside sources, making quantum leaps in technology and philosophy that can't be explained without intervention. You take the pyramid of Khufu – over two million stones with an average weight of five thousand pounds apiece, some as heavy as sixty thousand pounds. They had stone tools, like the Easter Islanders, and at that point had not yet discovered the wheel. But these aren't simple monoliths that could be hoisted up with some rope and a fulcrum. These are entire pyramids, taller than a modern forty-story building.

"Then the Mayans, who built great cities and grand astronomical observatories, also with stone tools, also without the wheel. They were able to predict eclipses of the sun and moon, and built city-walls and temples of stone without mortar, the stone cut so closely you can't slide a knife or a even piece of paper between them.

"The Toltecs and the Aztecs were instructed too. When Cortéz landed, Montezuma's people were expecting him, as foretold before the departure of Quetzalcoatl many centuries before.

"The rise of every early civilization, from Chile to China, shows evidence of contact. But it seems that each culture was allowed to decide on its own what the nature of the relationship would be. For the tribes of the great plains in northern America and Asia, it changed their religion. In China, South America, and the Middle East, it improved their technology. In central Africa, it affected the social structure. But in each case, there is an early intervention, then contact is cut off. It happened some four thousand years ago in the Mediterranean, and maybe a thousand years ago in Peru. Wherever people migrated, at one point or another, a society was chosen to receive help. Why they were chosen – who knows? Some of these cultures were peaceful, others extremely brutal and repressive. And why contact comes in stages, why it's not kept up continually, that's not understood either.

"Maybe the best theory is that we are involved in a sort of cosmic experiment. A little nudge over here, another nudge over there, and we'll see what happens."

"And when the whole planet went to war, with modern ships and planes..." I was beginning to see the tent take shape.

"They came out again."

"*Out*? Not *back*?"

"We've got no way to know. Carl Sagan said before he died that, even with today's science, we could design ships that would circle the known universe in just over fifty years at close to the speed of light. The only catch is, we're centuries away from being able to build them."

"Isn't there a problem with time when you do that?" I asked.

"Yes, there's that. The closer you get to the speed of light, the more time slows down. So if we were all to get into one of Carl's spaceships and cruise the known universe for fifty years, by the time we got back to Earth... well, there'd be no Earth. It would have been swallowed up by our sun's transformation into a 'red giant' star. From the Earth's point of view, millions of years would have passed."

"But how is that possible?"

"You've heard of $E=MC^2$."

"Energy equals matter times the speed of light squared."

"Exactly. What that means is that matter can be changed into energy, and vice versa, the way water can be changed into steam and then back into water. And just like you get a whole roomful of steam from a bucketful of water, you get a whole lot of energy from a little bit of matter. In fact, it's this explosion of matter into energy that fuels the atomic bomb.

"And this is why nothing can travel faster than the speed of light. At the speed of light, all matter has been changed into energy. There's no more water to make steam out of, if you will, so you can't go any faster."

"I don't see the connection," I said. The tent was sagging again. I was trying to hold onto too many poles at once.

"Bear with me. You've heard of the space-time continuum."

"Yes."

"That's another one of Einstein's discoveries. It turns out that space and time fit together like matter and energy."

"You mean we can make space into time and time into space?"

"That's exactly what's happening when astronauts travel so fast."

"How?"

"Did y'all drive up from the south?"

"Yeah." I decided to trust him on this one and just let

him go wherever he was going.

"Y'all passed that vegetable stand just down the road?"

"Uh-huh."

"Well, suppose I took a walk down to that vegetable stand. I guess it'd take me about twenty minutes. Now, let's suppose you leave to go from here to the fruit stand at the same time I do, and we both arrive at the same moment, only you've gone around the sun at close to the speed of light while I've been just walking down the road. My watch will say 3:20 when we get there. Yours will say 3:02. And they'll both be right.

"Because you used up more space, you wind up using less time. When astronauts come back from outer space, they're just a few seconds younger than the rest of us who stayed behind, because they've been going so fast. It happens when you fly in a plane too, only the difference is so small you can't even measure it."

"You mean, the faster you go in space, the slower you go in time."

"That's pretty much it. It all has to balance out."

"If that's true, then how could anyone contact us?" I demanded. "If time's going so much slower for them while they're travelling, their whole civilization will have vanished by the time they get back. And they would have had to leave their own star system before there was anything on Earth to know about."

"That would be true if space-time were flat," John said. "But it's not. It's curved."

"You mean like the Earth is curved?"

"That's what Einstein thought. But the reality turned out

to be much more complicated."

"The universe is complicated. But the idea is pretty simple. Take a look here." He held one end of his blanket stretched out between his two hands. "How long do you think it would take an ant to get from one of my thumbs to the other, walking along this blanket?"

"I don't know. Maybe a minute."

"How about now?" he asked – and he folded the blanket together so that his thumbs met.

"No time at all," I said.

"Exactly. Quantum physics records particles vanishing in one place and appearing simultaneously in other places. Our best guess is that they're taking hyperdimensional shortcuts across warps in space-time, where it bends back on itself like one of these hairpin curves on the mountain."

"First, let's look at what we know. We know someone has been introducing advancements in human civilization for thousands of years, at different times and places around the globe. We know that during World War II, alien craft were spotted worldwide, and are still being seen. We also know that close encounters, visitations, and abductions are on the rise. Now, add to that the discovery that particles can travel across warps in space-time, and that presumably it's possible for larger objects to do the same...."

"Wait!" I said, so loud that I was afraid I would wake up Ron and Karen. "If it's possible to travel through warps in *space*, and if space and *time* are interchangeable...."

"I think you're catching on," John said.

We sat quietly for a moment, listening to the rain, which

had eased up some, and to the cracking of the low fire. I expected John would get up to add another log, or to go breath some lightning, but he didn't move.

"Now I'm going to tell you something I haven't told anyone except my wife," John announced after a while. I said nothing. "My cousin swore me to secrecy while he was alive, but he's gone now. He was a colonel in the Air Force, won four Distinguished Flying Crosses in the War. He was out at Roswell Army Air Field in New Mexico in July of 1947, when the Air Force was called in to investigate that crash on Max Brazel's ranch between there and Albuquerque.

"If the brass had gotten there a little sooner, they might have cleaned it up and debriefed everyone and gotten their official version prepared and released, and no one would have known about it. But on July 8th, before things were too well organized, the base's public information officer, a Lieutenant Walter Haut, released a statement to the media that 'a flying saucer' had been recovered. The story was squelched, the military claimed it was a weather balloon, but then four alien bodies were discovered about thirty-five miles east-southeast of the crash site. I suppose the government thinks they were hitching a ride on that weather balloon.

"All of that's pretty well known by now. The government still denies it officially, but the Air Force Museum at Wright-Patterson Air Force Base in Ohio – that's where the bodies were sent – has its own 'UFO' display, including models of the alien corpses.

"What isn't known, not publicly at least, is that there was a sphere, a round ball highly polished. He said it must have been eight to ten inches in diameter. It was floating in mid-air. All he would tell me about it was that a diamond bit couldn't scratch the material, and that records of the sphere were purged just after it went back east for analysis.

"But it really doesn't matter what gets classified or covered up. Little by little, they're making themselves known. If I'm not mistaken, you've seen them."

I nodded.

"Well, I won't ask you for details."

"When did you first see them?" I asked.

"Depends on what you mean," he replied. "The first time I saw a craft was in 1953. I was driving out of Amarillo toward Mexico, and had crossed over into New Mexico from Texas, listening to a radio station out of Albuquerque, when the radio started scrambling and my car was trying to quit on me. I looked up to the east-southeast and saw something. It looked like a silver cigar, almost like a kind of balloon, but it wasn't a balloon. It had the color of a new nickel with a kind of satiny finish.

"I stopped the car and got my binoculars out to watch it. It was hanging about a hundred feet over the desert. All of a sudden, there were seven small craft coming toward the big craft. They all lined up right under the big one and one at a time they entered through a bay. I would guess the long craft was the size of two or three football fields. The disks were much, much smaller. After the little ones had all entered the big one, it took off toward the south-southeast, toward Mexico. It was moving kind of slow. The next day it was in the paper, so it can be verified. It was near a little town. I'd have to look on a map to tell you the name. You don't happen to have a map with you, do you?"

"Not of New Mexico."

"Now," John continued, "let me tell you about a thing – one of many things – that we witnessed in the Triangle. We were headed down to pick up a group of people who had flown in from Italy. We were on our way to meet them in the Bahamas. All of a sudden I said to Karen, 'I hear something like two trains running together.' She said, 'Yes, I hear it too.' We were out in the open ocean at the time. The loud roaring sound kept getting louder and louder. So she stepped out to the port side of the bridge and I heard her say, 'My God, honey! Put it on automatic pilot and come look at this.' We saw what looked like two giant waterspouts running together. It made the worst sound you ever heard and they were coming right at us. I kicked the throttle wide open and made a ninety degree turn. They went right by us. It looked like that each one was about a hundred yards across. If they had hit us they would have sucked the boat under water. Well, they got to a certain point and went under the water with huge force. We could see in the middle of the spout there was something shiny and silver and it took it straight down below us.

"At a later date, I was with several scientists and we were up on the bridge. We had been studying an underwater roadway from an eighty-foot oceanographic research ship called *The Dragon Lady*. We were coming from Bimini and headed toward Ft. Lauderdale. Suddenly, my automatic pilot wasn't working. It started to get harder to pull and the vessel came on back around. So I reached and pulled one of the throttles out on the diesel. No outside interference would bother a diesel. It runs on compression. You can take the battery out of a diesel and it will keep on running.

"I pulled it back to neutral and we were just lying in the water. Dr. Valentine said, 'Good Lord, what's this?' We had a nine-inch radar scope. On its grid we saw a huge thing coming very slow over the water."

"Could you see it?"

"Before too long, yes. It was about fifty feet above the water, saucer-shaped with a dome on top of it. I saw three port lights. They had a dull reddish glow around the perimeter. Dancing around it was a greenish-blue light. It was like Saint Elmo's fire and it was going counter-clockwise. The ship was a silver color with a satin finish – like the color of the ship I saw near Mexico – and it had strange markings on it."

"John, if you saw those markings again, would you recognize them?"

"I sure would. It's implanted in my brain. But I've seen them before. Anyway, you asked me a while back what the War and the out-of-towners had to do with the Bermuda Triangle, and that's what I'm getting at."

Chapter Seven
Rumbles from Atlantis

"The reason we were out in the *Dragon Lady* to begin with was because of what Dr. Mason Valentine found off the north island of Bimini in 'sixty-eight. It's called the 'Bimini Road'."

It was in September of 1968, to be precise, when Dr. Valentine of the University of Miami discovered the underwater roadway. About three thousand feet from shore (in his words) "I was amazed to discern an extensive pavement of rectangles and occasionally polygonal flat stones of varying size and thickness, obviously arranged and accurately aligned to form convincing engineering courses. These stones had evidently been submerged over a long span of time, for the edges of some had become rounded, giving the blocks the appearance of giant loaves of bread, or pillows. My personal feeling is that the whole fantastic complex represents the intelligent utilization by ancient man of material provided by nature and appropriate for the creation of some sort of ceremonial center."

The hundred-yard-wide stone pathways (there are two of them, running parallel) do not follow the curve of the shoreline, as reefs do, but run straight for some 1,600 feet. Many of the stones are flint-hard micrite, and some of the larger ones are flanked by the ruins of pillars.

In 1989, Dimitri Rebikoff, inventor of the Pegasus

underwater platform, discovered a five-hundred-foot-long subaquatic mound in the shape of a fish during a fly-over of the Bimini Road site. Shortly thereafter, an aerial surveying firm out of Miami was hired to take infrared aerial photographs of the Bimini area at various levels of altitude. These not only established the east-west orientation of the fin at the fish mound, but also revealed two other mounds – one rectangular, and another in the figure of a cat. Evidence is starting to appear that this section of road is probably an especially well-preserved portion of a much larger structure reaching through and beyond the islands.

"The funny thing about the Bimini Road being discovered in 'sixty-eight," John continued, "is that Edgar Cayce predicted that parts of Atlantis would be uncovered in 'sixty-eight or 'sixty-nine."

"You think this road might be part of the lost continent?"

"Well, from what records we have, it's estimated that the continent itself suffered some sort of cataclysm, maybe an earthquake or a volcanic eruption, over fifty thousand years ago. Some twenty thousand years later, another disaster hit the continent that broke it up into three islands, which the ancient Greeks called Aryan, Og, and Poseidia. These islands were inhabited, according to Plato, until about the year 10,000 BCE, when something caused them to sink.

"Chances are, the islands didn't just fall off into the ocean all at once. Islands are always forming and sinking in the oceans, but it happens on a geological time scale, and that's a lot longer than your average human's three-score-and-ten years."

"What could make an island sink?"

"Any number of things, really. Small, level islands can actually erode away. But that's not likely for Atlantis, because we know the Atlanteans were advanced enough to build cities and carry on wars, so I doubt they'd just let their islands wash out from under them. Unless, of course, a tsunami overwhelmed them. Or, it could have been a fissure opening up in the seabed from some sort of volcanic pressure. Could have been the result of plate techtonics. Earthquakes – maybe several of them over a period of years – might have done it. Then again, if the bedrock of the island were porous, the foundation might have become unstable, sort of like a sinkhole on land. It's possible that the Bimini Islands are all that's left of the catastrophe, the tops of some Atlantean mountain peaks.

"Whatever it was, it seemed to happen slowly, or in stages, since a good portion of the population emigrated to Egypt in the years before the islands disappeared. An Egyptian High Priest of Atlantean origin, named Ra Ta, had some positive influences on Egyptian culture, as far as human rights and philosophy. The ancient Egyptians, you know, were more open than we are these days about letting foreigners get to high positions in national government – Kissinger being an exception, I guess. If you remember, Joseph – a Hebrew – was second in command to Pharaoh himself. It says in Genesis, 'And Pharaoh said to Joseph, "Behold, I have set you over all the land of Egypt." Then Pharaoh took his signet ring from his hand and put it on Joseph's hand, and arrayed him in garments of fine linen, and put a gold chain about his neck; and he made him to ride in his second chariot; and they cried before him, "Bow the knee!" Thus he set him over all the land of Egypt.'

And of course you know about Moses. It's very likely that some of the priests of the Atlantean tradition helped these two, since they also worshipped the One God like the Hebrews did, and didn't bow down to idols.

"But the Atlantean philosophy didn't become official until the empire of Pharaoh Akhenaten in the late Bronze Age. He moved the capital and established the worship of the True God. But after he died, the pagan pharaohs took over again, moved the capital back to where it used to be, and started worshipping all the pagan gods and goddesses and ordering everyone else to do the same. Just like the early Christians were to Rome, the Atlanteans were a threat to the powerful in Egypt. Unfortunately for them, Akhenaten didn't have the staying-power that Emperor Constantine did. It was Constantine that finally converted the Roman Empire to Christianity. But it was just too much for poor old Akhenaten in Egypt.

"As far as we know, the ancestors of the Hebrews and the ancient Atlanteans were the only people in those days to reject paganism and accept the one and only God. And now the State of Israel has been re-established on the face of the Earth, as was prophesied. That came out of the War too. And according to Edgar Cayce, who was right on the money about the Bimini Road being discovered in 'sixty-eight, more of Atlantis will resurface, and parts of the land that's above water now, even in the United States, will somehow be submerged into the ocean sometime over the next several decades. Maybe that earthquake will finally hit California. Maybe global warming will raise the sea levels like the scientists have been saying it will."

"Do you think the Atlanteans will return?"

"Some of them are already here. Some of them never went away. You and me might have Atlantean blood in us for all we know."

"Do you think they were visited?"

"Almost certainly. In fact, as one of the oldest advanced civilizations on record – remember, they were destroyed before most cultures had even arisen – they might have been the first people contacted and instructed on this planet. According to Plato, they possessed the most developed technology of the day. They lived in peace and prosperity. But then, something changed. Maybe their own source of knowledge abandoned them at this time. Their culture degenerated. Crime and corruption increased. They began to wage wars, but were eventually defeated by the Greeks and withdrew to the islands. An all too familiar story."

At that point John fell silent. We sat in the dimness for several minutes, not speaking. Finally, John got up to rekindle the fire in the fireplace, and to put another blanket over Karen, who looked curled up from the chill. When he came back, I said nothing, waiting for him to speak again. Since he hadn't gone outside, I figured there was something more he had to express.

"You know, Alex," he said at last, "we have to be careful. Just look back over history. The Altanteans were the richest people on the Earth in their day. They obviously had some sort of tremendous power source, possibly even nuclear (as Cayce concluded), which may have hastened their own destruction. And now they're gone. The Romans too – they ruled Europe and the Mediterranean. But they got too wealthy, too powerful. And the Egyptians and the Mayans with all their fine monuments, and the Easter Islanders as well. Like the Word of God says in the Book of Revelation, 'For you say, "I am rich, I have prospered, and I need nothing"; not knowing that you are wretched, pitiable, poor, blind, and naked.'

"And we're on the brink of it now ourselves. We've got more money and more luxury and more... more *stuff* than any civilization since the beginning. And we've also got more military power. Only this time, we've got the power to take the whole world out with us. For the first time in history, a single war could wipe out every living being on this planet, plant and animal. And it was World War Two that unleashed that power.

"I cannot believe – and I do not believe – that it is simply a matter of coincidence that Atlantis is once again being uncovered in these days, and that the, um, the out-of-towners we'll say, are more active than they've been since ancient times."

"So what about those disappearances in the Bermuda Triangle?" I interjected.

"Something is moving in and out of that space. Do you know why they move the lifeboats away from a sinking ship as fast as they can?"

"So the ship doesn't suck them down with it."

"Exactly. Anything of any size entering the surface of the ocean creates an enormous suction. There are all kinds of cloaking devices that might be used for airships or spacecraft – who knows what a more advanced culture might have discovered or invented? – and they might even be able to take advantage of warps in space-time to travel great distances, but one way or another something's moving in and out of the ocean in the Atlantic for some reason. And we can be pretty sure it's not anything made on this Earth. At least, not these days."

"You're coming back to that again," I said, remembering what he had said about space and time being two parts of the same thing.

"That's right, I am. There's just no good way for us to know, until and unless we're shown more than we've been allowed to see so far, whether our visitors are travelling in space or in time. They may be our own people after all. Then again, maybe not. But one thing's clear, they're taking a lot of risks, probably showing more of themselves than they'd rather. They're in a hurry, Alex. Something's got to be done for this planet, and soon."

"You know something, John – I feel like I don't really need to tell you this – but I'm sure that they're here to help."

"Yes indeed, I am too, little lady. There's no doubt in my mind (and I'll let you make of that what you will) that those missing pilots and crewmembers of all those boats and planes are alive and well, and we'll see them again. And they won't be very much older than when they disappeared, even the ones from fifty years ago. And they will know a lot. They're likely to be leaders in this world sometime after the Millenium. People like you and me, we're just here to prepare the way, to ease the transition, and there are plenty of others like us. Let's just hope there's enough."

Chapter Eight
Millenium

As John and I approached the wee morning hours before dawn, I wanted to know what he believed we were facing as we moved toward the Millenium, with so many preachers and psychics announcing the Final Days, and so many scholars and environmentalists warning of impending disasters both natural and manmade. In these days of grim hope, it can be difficult to hold onto our faith in salvation, in the midst of fire and flood, of wars and rumors of wars. Jesus said, "Nation will rise against nation, and kingdom against kingdom, and there will be famines and earthquakes in various places; all this is but the beginning of the birth pangs." I asked John about this, about how he felt on probably the most frightening and most important issue of the modern epoch.

"I do believe we can save ourselves," he confessed, "if we follow our faith and let God guide and direct us. It's in the Lord's hands after all, we can't decide it ourselves. 'Of that day and hour no one knows, not even the angels of heaven, nor the Son, but the Father only.' That's from the Gospel of Matthew. Jesus says that it will be like the days before Noah, when 'they were eating and drinking, marrying and giving in marriage, until the day when Noah entered the ark, and they did not know until the flood came and swept them all away'.

"So maybe this is one thing and maybe it's another. We'll have to wait and see. But whatever's happening, it doesn't just involve this world – it's a lot bigger than that. There's a cosmic turning point ahead.

"I'm glad I'm a man of the sea. If the seas rise, I'll be right at home. I'll sail my boat right up to the mountaintop. Isaiah said that in the Day of the Lord, the faithful should flee to the mountaintops, and live there. And Jesus repeated the words of that prophet to Matthew at the Mount of Olives, saying that when the sacrilege prophesied by Daniel came to pass, 'then let those are in Judea flee to the mountains'. And Isaiah said too, 'It shall come to pass in the latter days that the mountain of the house of the Lord shall be established as the highest of mountains, and shall be raised above the hills; and all the nations shall flow to it, and many peoples shall come.... and they shall beat their swords into plowshares, and their spears into pruning hooks.'

"I believe that the mountain ranges, like the Rockies and the Appalachians, like the Andes or Himalayas or Alps are all safe havens. And you want to stay clear of fault lines. Get a geological map and learn where they are. If you aren't near the mountains, get up into the hills. In the old days, caves represented safe havens. And I believe that many of the houses of God will be spared. But the only thing that will protect us in the End Times is serious prayer."

"I think we might be there, John. I've been getting some strong messages. One afternoon, I woke from a nap to find I'd been given a message to start stocking up on food and water. A couple of days later, I had the same dream or encounter. A few nights later, I was spoken to a third time, and encouraged to get prepared and stock up on plenty of water, now, don't delay."

"Alex, you had better do just that. End Times or not, it's going to be a bumpy road ahead. But if these are the Latter Days, those of us who are survivors will have the Garden of Eden right here. It says in the Book of Revelation, 'I saw a new heaven and a new earth; for the first heaven and the first earth had passed away, and the sea was no more.' The old sky and the old Earth and the old sea will be gone. It'll be a new world, where God will be with his children again, like before the Fall. In his vision at Patmos, John says 'I heard a loud voice from the throne saying, "Behold, the dwelling of God is with men. He will dwell with them, and they shall be his people, and God himself will be with them; he will wipe away every tear from their eyes, and death shall be no more, neither shall there be mourning nor crying nor pain any more, for the former things have passed away".'

"It could mean that Atlantis will rise again. Maybe some of the first people to worship the One God will return for that day. There's a lot in store. Some of it has been given to me in dreams too. Like I said, my father was Cheyenne, and the divine gift of dreaming is very strong with that people. I've seen thousands standing in the muddy waters washing, in the healing mud of the volcanoes. Maybe these volcanoes will be the ones God uses to create the new Earth.

"And I've seen that the seeds held for centuries in the pyramids of Egypt will be planted again, and they will flourish all over the world."

Oh, if you'd been able to hear my heart when he said

those words. I swear it skipped a beat, maybe even a beat and a half. I latched onto John's arm without knowing, and as dreamy as he'd become there in the early hours leaning back on the sofa he startled like a colt.

"You've seen them too?" I said in amazement.

John just shook his head and grinned. "You scared me for a minute," he said. "I thought it was something serious." And then he laughed till he couldn't talk almost for a minute, and got me laughing too. We were like schoolkids all of a sudden, trying not to wake the others with our snickering, which only made it harder not to giggle. I think I said already that once I get going I'm hard to stop.

"No, John," I said when I got my breath, "I *am* serious... I mean I'm... I'm for real, John...."

He squeezed my arm with his fingertips like he was sampling a loaf of bread. "Yes, I can see that," he said, "you're for real all right." He was getting silly. For a moment I was frustrated, almost angry, but then I caught the star-sparkle in his eye and the message came through. Now he took my arm in his hand, like a father would.

"You've got to accept it, Alex," he told me. "Just let it be."

"But a woman named Celia, I never even knew her last name, she had these seeds, in a matchbox, they were from the pyramids and she'd given up all she had to protect them. I don't know where she was from, or where she was taking them...."

"Are you really surprised, Alex, that she would find you and confide in you, and that you would find me, and I would be able to confide in you, get part of my story told? Do you think things happen for no reason? Of course not. You and I both know there's a cause behind it all. You're going to have to accept your role in all this, just like I am and everyone else is. If you're a messenger, you take the messages you're given, and you deliver them where you're told. That's all you can do."

That's all you can do. I suppose that's true.

Chapter Nine
Shiny Black Shoes and Angel Wings

"It's gonna be morning soon," John observed. Suddenly I was exhausted. The storm had calmed into a mountain rain. Outside the windows was utter blackness. I thought how the room had looked to me only a few hours ago, when we quite literally stumbled into it. How subaquatic and aquarium-like it had appeared. Now it seemed far and high above the low-lying waters of the world with their drastic fishy depths. We were up on the mountain, raised into the air, almost past the storm and night.

But weary from it now. All this talk and thought. A jealousy came over me for Ron's night of dreamy sleep . But what I knew now was worth a hundred nights' rest.

"I think you'd better get yourself a couple hours of sleep, little lady," I heard John tell me. I must have had my eyes shut. "I'm going outside and breathe some lightning."

"Aren't you tired?"

"I usually don't get to bed till late anyway, and then it takes me till sunrise to get to sleep, and then it's time to get up again."

"I want to go out too, before I go to bed."

The rain smelled sweet by now, even through the cold. Batting against the leaves, it sounded like thousands of tiny gnomish horses stampeding through the sky. We sat in the rockers in silent watchingness a while with our blankets around us like swaddling clothes.

"Tell me one more story, John, and then I'll turn in."

"I'll tell you maybe the strangest one I know. It happened to me when I was a boy, in a county called Anderson, not far from the Louisiana line. My aunt had just bought an old house, way back down a road from nowhere, and I was to meet my mother there to help out with things that needed doing. I remember that front door, it was rounded at the top, solid wood, wide as a barn door, and felt twice as heavy. I walked in and called for my mother, thinking she would be there.

"The leaded crystal windows stretched almost to the ceiling, and the floors were hardwood. The house really had a cold feel to it. As I walked through, I got the feeling that someone was watching. I kept turning to see, but no one was there. I called my mother. All I could hear was the birds and the squirrels fussing out in the trees.

"I came to a hallway. It ran crosswise through the house, cutting it in two, so that the one side was built as a kind of mirror image of the other side on that floor. I heard someone call my name. Call me by my own name. It felt like a drift of snow had fallen on my body and spirit. There stood a man in a black suit (that's what I thought of it – I'd never seen a tuxedo before) right across the hallway. A fine-looking sort of man. He had high-buttoned shiny patent leather shoes on. Till the day I die I will remember those shoes.

"He smiled the biggest smile, and I saw white teeth that were perfect. I'd never seen perfect teeth before either. I mean they were so perfect they looked carved out of ivory. But his hair was jet black, slicked back very neatly; it reminded me of a raven's neck, the way it'll show color in the light it's such a deep dark. He raised an eyebrow at me and I felt myself shiver. His eyes were so black you couldn't find a pupil in them.

"He stuck his finger out and motioned for me and said with a clear voice, 'Come here, son'. Like I said, he was a fine-looking sort of man. Every feature fit just so. But there was cruelty in the arrangement. I can't explain it. Something told me, don't go near him; he can't hurt you if you don't go near him. So I stayed put, as well as I could with my legs losing their bonyness underneath me.

"Again he said, 'Come here. I'm not going to hurt you. Come here.' I said, 'No sir. I know who you are.'

"Here's the strange thing. When he got ready to leave, he was gone. He turned and was just gone. The hallway smelt like sulfur.

"You know if you are on the right side when you are visited by the Dark One. And we are all tested. I just don't know why I was tested so young."

"John," I said, "I can't go to sleep with the Dark One on my mind. You owe me another story."

"All right. I remember when I was about six years old, I had a dream one night. I call it a dream, but I can tell you it was real enough and I wasn't asleep. A greenish blue light came into my room and I heard music. It was music like I'd never heard before. It was just beautiful, and I saw all the shapes floating – but it wasn't all physical, so I can't really tell you. And I was small, I didn't think about having to put words to it. It only lasted a moment, but it is very distinct. And several years later I heard that music again."

I don't remember walking back to my cot, but I do recall lying there, and the fire, and thinking, "I wonder where Ray and Kitty are, if they made it home all right." I didn't doubt that they had.

Afterword

"My Flag"

It's a nice thing to be woken up by laughter. It doesn't happen often enough. Hardly ever, when you think about it. But I was awoke by laughter, and there was the smell of coffee and buttermilk pancakes. It was late morning.

A couple of truck drivers were sitting at two tables along the wall. They nodded politely when I stuck my head up from the cot. One of them held up his coffee cup a little toward me, like some sort of good-morning toast to waking-up-ness. Since I was still in my clothes, I just stood on my feet and went to have a cup of coffee with them. Their names were Billy and Calvin, and one was as big as the other was skinny, and they'd met last night at the emergency turnoff and sat out the storm playing cards and telling lies in the cab of Billy's rig.

"Your people's all in the kitchen," Cal told me. He was the small one.

"We done had breakfast a-*while* back," said Billy. "We were just waiting to see how long you would sleep."

I had fixed myself a cup of coffee from the Bunn-O-Matic automatic coffee pot, and I sat and talked with the two drivers until it was finished, and then I fixed myself a second cup on the way to the kitchen. Everyone applauded when I entered, so I took the curtsey. I must have looked a fright. Pat had kept some pancakes in a warmer and some syrup in a double-boiler on the stove for me. She said Jay-El was outside pretending to clean up storm damage.

I leaned against Ron like he was a tree. He smelled of soap and his hair was washed. He kissed me on top of my head and said good morning. No telling how long he'd been up. Karen too, she looked like an early riser. John simply hadn't gone to sleep, I don't suppose.

"John says you wrote a book about him last night," Ron said.

"I guess I did," I yawned.

"What's going on the cover?" John wanted to know.

"The American flag," I said, right off the top of my head.

"Well, then you get one more story," John replied. "I remember one day I was on leave in Honolulu, Hawaii. In uniform, all by myself, walking along the beach. Then I start hearing this sound, like a *sloosh slish slush*, and some voices. I turn around, there's fifteen hippies dragging an American flag down the beach. All fifteen of them, dragging one flag, each one with a hand on it.

"I yelled 'Hey there!'

"They all yelled 'Hey there!' back at me. They were stoned halfway to next Friday, and a little drunk on the side I think, some of them. I hollered, 'What are you doing with my flag?'

"This one fellow pipes up, 'We're dragging it. We're gonna s--- in the ocean, then wipe our a--es with it, then burn it!' I don't know if that was really their intent, or if this kid had just made it up on the spot to get my goat. But they were dragging it at least, and a few of them seemed receptive to the other three suggestions.

"I'd turned toward them and we were close enough to speak without shouting at each other. I said, 'No. You're not.'

"'Well,' this fellow says, 'we vote to burn it, and it looks like you're outnumbered.'

"'Could be,' I said. 'But if you don't give me that flag, I won't be nearly as outnumbered in a few minutes as I am now.'

"They thought about that for a moment. And after they'd given it due consideration, I folded up my flag and walked on.

"I've still got that flag."